PHOTOSYNTHESIS

Regulation Under Varying Light Regimes

V.S. Rama Das
INSA Honorary Scientist
Visiting Professor, Biotechnology Centre
Nagarjuna University
Nagarjuna Nagar
INDIA

Science Publishers, Inc.
Enfield (NH), USA   Plymouth, UK

CIP data will be provided on request.

SCIENCE PUBLISHERS, INC.
Post Office Box 699
Enfield, New Hampshire 03748
United States of America

Internet site: *http://www.scipub.net*

sales@scipub.net (marketing department)
editor@scipub.net (editorial department)
info@scipub.net (for all other enquiries)

ISBN   1-57808-343-5

© 2004, Copyright Reserved

All rights reserved. No part of this publication may be reproduced, stored in a retrieval system, or transmitted in any form or by any means, electronic, mechanical, photocopying or otherwise, without the prior permission of the copyright owner. Application for such permission, with a statement of the purpose and extent of the reproduction, should be addressed to the publisher.

Published by Science Publishers, Inc., Enfield, NH, USA
Printed in India.

Preface

Light quantity — apart from quality a crucial environmental factor determining the photosynthetic performance and plant productivity in view of the fact that under natural conditions, it is highly variable over several orders of magnitude even during the course of the day. Plants have evolved multiple strategies to mitigate the adverse effects of such often sudden exposure to varied levels of irradiance for different periods of duration as also to optimize the rate photosynthesis at a given light intensity. Therefore, the regulation of photosynthesis by the level of incident light and the amount of energy harvested is an area of paramount importance in photosynthesis research.

The aim of this monograph is to highlight the available knowledge of the various mechanisms followed by plants to overcome the high light stress and photoinhibition, as also the plant responses to low light levels. Emphasis has been on the regulation of light harvesting process, the avoidance phenomena at leaf and chloroplast levels and the strategy followed by certain solar tracking plants in maximizing the light interception without photoinhibitory damage, leading to improvement of diurnal photosynthetic rate. Plant adjustment to light variations over a long term through the acclimation process brought about by changes in gene function and specific reconstruction of chloroplast has also been discussed.

This text attempts to present a broad view of the research in the field of light regulation of photosynthesis providing a framework rather than an exhaustive account of the discipline. A detailed analysis of all the available data has not been possible in a monograph of this size but is intended to produce an awareness of the latest trends of research by extensive citation to current literature and also to identify the gaps existing in the current knowledge. Throughout the book , the predominant approach has been on higher plant systems, though green algae and cyanobacterial examples are used wherever appropriate.

It is also assumed that the reader has sufficient background knowledge of photosynthesis. The information presented in the text has been documented predominantly based on original publications, but at some places, certain recent critical reviews have been cited for their informative value.

The treatment of the subject has been based on a sequence of chapterization as follows : Chapter 1 deals with the basics of light harvesting process and the organization and structure of light harvesting antennae systems in order to appreciate the regulatory mechanisms described later. Very briefly, the structure and organization of the photosystems II and I has been presented to provide an understanding of the structural changes in the chloroplast apparatus due to varied levels irradiance mentioned in the subsequent chapters. Chapter 2 covers the issue of photoinhibition, a phenomenon of central importance in photosynthesis and is of considerable practical significance. The events of photoinactivation, degradation of D1 protein and recovery of functional photosystem II are presented in detail.

After a discussion of photoinhibition, it is logical to understand the meaning of photoprotection and the mechanisms involved to overcome stress. Chapter 3 deals with strategies known to operate in plants, particularly against excessive light energy absorbed. Thermal dissipation of energy, xanthophyll cycle and the scavenging of reactive oxygen species are the principal areas covered in this chapter. Photoprotection is currently an active field of research and, therefore, some of the relevant areas are presented in a little more detail.

Plants have evolved mechanisms for controlled absorption of light energy, particularly through adjustment of leaf orientation to solar beam. Leaf heliotropism is the subject matter of Chapter 4. The leaf movements related to the principle of light avoidance, the paraheliotropism and the opposite strategy, diaheliotropism leading to maximization of light absorption have also been discussed. Apart from short-term effects of light stress, the adjustment of plants for long-term acclimation to varied light environment is a matter of considerable importance.

Chapter 5 emphasizes the process of acclimation of photosynthesis to the light environment and describes the relatively long-term changes in structural attributes of chloroplasts and related aspects. Chapter 6 provides a short account of the use of transgenics and of biotechnological approaches in research pertaining to theme of the monograph. In the last chapter, certain existing gaps in the present knowledge in the area of light regulation of photosynthesis and possible future lines of research are highlighted in the form of concluding remarks.

Enough care has been bestowed in the citation of all relevant literature available in the area of regulation of photosynthesis in a dynamic light environment. However, it is possible that the work of some authors might not have been included by oversight, which is regretted. The author shall be grateful to the readers for bringing his attention to any errors or omission in the text. Finally, it is hoped that this modest effort of a monograph would be an useful reference to advanced postgraduate students, research scientists and teachers interested in the area of photosynthesis research.

This monograph is prepared with a financial aid from the Department of Science and Technology, Government of India as a project sanctioned to the author under the Utilization of Scientific Expertise of Retired Scientists (USERS) Programme and it is gratefully acknowledged.

Several former associates of the author have provided library materials as and when requested. In this respect my thanks are due to Dr. Veeranjaneyulu Konka, Dr. T.V.S. Sresty, Dr. I.M. Rao, Shri J.K. Prahar, Dr. P. Prasanth and Prof. A.S. Raghavendra for their timely help. I thank Prof. S.V. Sharma for his sustained technical help during the preparation of this work. I also thank the authorities of Nagarjuna University for having provided the opportunity to take up this project. My thanks are due to Shri M. Srinivasulu Reddy for typing the several drafts and the final copy of the manuscript. I thank my wife Smt. V. Ahalya Devi for her continued encouragement.

Contents

Preface iii

1. Light Interception : Basics and Structural Aspects 1
1.1. Introduction 1
1.2. Light Harvesting 2
1.3. Structure and Role of Antennae 4
1.4. Composition of Photosystems 12
1.5. Photosystem II : Composition and Structure 14
1.5A. Reaction Centre of Photosystem II 17
1.5B. Core Complex of Photosystem II 19
1.6. Photosystem I – Structure and Organization 19
1.6A. Reaction Centre of Photosystem I 20
1.6B. Core Complex of Photosystem I 23

2. Photoinhibition 26
2.1. Introduction 26
2.2. Occurrence of Photoinhibition 28
2.3. Photoinhibition of Photosystem II 29
2.3A. Acceptor Side Photoinhibition 30
2.3B. Donor Side Mechanism 32
2.4. Photoinactivation 33
2.5. Degradation of D1 Protein 35
2.5A. Proteolysis of Damaged D1 Protein 36
2.5B. Resynthesis of D1 Protein and Insertion 40
2.5C. Restoration of Functional Assembly of Photosystem II .. 43
2.6. Photoinhibition of PS I 45

3. Photoprotection 47
3.1. Introduction 47
3.2. Thermal Dissipation of Excitation Energy 51
3.2A. Site of quenching 56
3.2B. Mechanism of qE 57
3.3. Xanthophyll Cycle 61
3.3A. Mechanism 62
3.3B. Pigments of Xanthophyll cycle and their Structure .. 66
3.3C. Role of Xanthophyll Cycle 68

	3.3D. Prevention of Lipid Oxidation Stress	70
	3.4A. A Reactive Oxygen Species (ROS) : Production of Active Oxygen	71
	3.4B. Antioxidant System in Scavenging Active Oxygen	72
	3.5. Additional Electron Sinks	79

4. Leaf Heliotropism, Solar Tracking and Regulation of Light Interception — 84
    4.1. Introduction — 84
    4.2. Paraheliotropism and Significance — 86
    4.3. Diaheliotropism — 87
    4.4. Leaf Heliotropism and Photosystem II Efficiency — 87
    4.5. Site of Perception and Mechanism of Leaf Movements — 92

5. Acclimation of Photosynthesis to Light Environment — 94
    5.1. Introduction — 94
    5.2. Regulation — 95
    5.3. Acclimation to Irradiance Levels — 96
    5.4. Acclimation in Mature Leaves — 98
    5.5. Acclimation to Changing Light Regime — 101

6. Transgenic and Biotechnological Approaches — 106
    6.1. Introduction — 106
    6.2. Photosystems : Reaction Centres — 106
    6.3. Light Harvesting Antennae — 109
    6.4. Photoinhibition : Transgenics — 111
    6.5. Functional Genomics of Plant Photosynthesis — 115

7. Concluding Remarks — 119

8. References — 122

Author Index — 167
Subject Index — 174

1
Light Interception : Basics and Structural Aspects

1.1. INTRODUCTION

Light energy is essential for plant life to drive its photoautotrophic mode of nutrition. In nature, solar radiation is the prime environmental resource for photosynthesis as also the most variable environmental factor (Long et al. 1994, Alves et al., 2002). An important light parameter affecting photosynthetic performance is its quantity (Baker, 1996), and the usual expression for the quantity of sunlight is its intensity. It should actually be irradiance, which refers to the quantity of light energy incident on a surface and is measured in units of power, watts per square metre (Wm^{-2}). Photosynthesis is known to be more dependent on the number of photons (quanta) striking a surface rather than on the actual energy content of these photons. Accordingly, the photosynthetic irradiance can be expressed as the number of photons falling on a unit surface in a unit time. This expression on a quantum basis is referred to as Photosynthetic Photon Flux Density (PPFD).

Solar radiant energy comprises different wavelengths of light and the range of wavelengths utilized for photosynthesis happens to be from 400 to 700 nm. This part of the spectrum is appropriately called the photosynthetically-active radiation (PAR) and comprises approximately only 50 per cent of sunlight. PAR may be quantified either on the basis of energy units (Wm^{-2}) or in terms of the number of quanta, i.e. mole m^{-2} s^{-1}. The PAR of direct sunlight at its maximum measures around 2000 µ mol m^{-2} s^{-1}, which corresponds to 400 W m^{-2} in energy (Bell and Rose, 1981). The light energy should first be absorbed by the pigment system in order to be effective in photosynthesis. The energy of sunlight is captured by pigment systems, predominantly chlorophylls and additionally by carotenoids (Ke, 2000a). It is now obvious that there are two types of pigments located in the pigment protein complexes of the two photosystems in the context of oxygenic photosynthesis. These photosystems are integral membrane complexes of thylakoids in the chloroplast. The two functionally different classes of pigments include light harvesting but photochemically inactive antennae systems and the photochemically active reaction centre pigments (Hillier and Babcock, 2001).

1.2. LIGHT HARVESTING

The process of absorption of light energy and its subsequent transfer to reaction centres of photosystems is referred to as light harvesting (Horton et al., 1996). The light absorption takes place through chlorophyll and carotenoid molecules, which are bound to light harvesting complexes (LHC proteins) located in the thylakoid membranes which then results in singlet state excitation of pigments (Paulsen, 1995); Pichersky and Jansson, 1996; Simpson and Knoetzel 1996; Melis, 1996; Formaggio et al., 2001). The light energy utilization for photosynthesis involves the sequence of reactions of light harvesting and energy conversion through primary photochemistry at two different sites, the antennae system and reaction centers, respectively. Since the function of harvesting of light energy resides in the light harvesting system, the regulation of the amount of light absorbed depends on the antennae system rather than on the efficiency of transfer process to reaction centre. Thus, the light energy absorption and conservation in the form of chlorophyll excited states represents the initial reaction of photosysnthesis (Owens, 1996). The light harvesting systems serve the antennae function of absorption by increasing the cross-sectional area involved in the light absorption by photosystems. This can be brought about either through an increase in the number of pigment molecules associated with the reaction centre or by using pigments other than those located in the reaction centre and broaden the overall absorption band, thus increasing the efficiency of light absorption (Morishige and Dreyfuss, 1998).

In natural environments, rapid fluctuations occur in light intensity diurnally as well as seasonally (Niyogi, 2000). Plants have to adjust to changes both in intensity and quality of light that they encounter. Also, there is an excess of PS I capacity over that of PS II (Albertsson, 2001). At times, excessive stimulation of one of the two photosystems results in an unequal utilization of light and causes damage. These qualitative changes are encountered by higher plants, which apparently have evolved mechanisms to mitigate the inappropriate light energy distribution between photosystem I and II. This process of adjustment of the capacity of light absorption between the two photosystems is referred to as state transition, which is a rapid regulatory mechanism (Snyders and Kohorn, 2001). However, for the survival of plants, it is more crucial for development of mechanisms to adjust to variations in light intensities.

The quantity of light energy absorbed by plants in excess of that required for photosynthetic process is known as Excess Excitation Energy, (EEE) (Huner et al., 1998; Havaux et al., 2000; Mullineaux and Karpinski, 2002). An efficient transfer of excitation energy takes place from the light harvesting system of PS II to the reaction centre with a maximum quantum yield of electron transfer specifically under conditions of limiting light environment (Demmig – Adams and Adams, 1992; Horton et al., 1999; Ort 2001). As long as the light intensity remains relatively lower, high quantum yields are noticed while light saturation of photosynthetic yield occurs with increasing light levels. A light saturated rate of photosynthesis (Pmax) is attained with an increase in the growth irradiance. The absorption of light continues by the light harvesting system even after attainment of light saturation of photosynthesis, leading to an excess excitation energy in the pigment system, causing potential damage (Horton et al., 1999). Therefore, the light energy below the levels of saturation is actually utilized for photosynthesis, while the light that is absorbed above the saturation level results in an excess. A generalized concept of the rate of light harvesting, light utilization and the excess excitation energy is shown in Figure 1.1. Excess light absorption could occur either as a result of increased absorption of incident intensity or due to lowered photosynthetic performance arising from environmental stresses (Owens, 1996). The excess light, therefore, is dependent on its existing environmental conditions under different levels of irradiance (Ort, 2001).

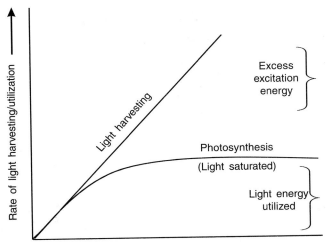

Fig. 1.1. A generalised diagram to show the light intensity dependence of photosynthesis. With increased light intensity, the rate of photosynthesis attains saturation while the absorption of light energy by the antennae system continues beyond that utilized for light saturated photosynthesis resulting in excess excitation energy.

Plant life obviously has to maintain a balance between efficient light harvesting under light limiting conditions and the regulated dissipation of energy under excess light conditions (Owens, 1996; Demmig-Adams and Adams, 2002). It is considered that plants have evolved a refinement of photosynthetic apparatus suited for high efficiency under limiting light and with critical regulatory mechanisms for performance with minimum damage under conditions of high and excessive intensities (Ort, 2001).

An understanding of the structural organization and the functional relationships of the light harvesting systems is essential to fully appreciate the precise mechanisms involved in the adjustment of plants to the inevitable exposure to variable light levels. Since LHC is a component of the holocomplex of the photosystems, it is also pertinent to consider the existing knowledge of the structure of the photosystems and their reaction centres. The process of photosynthesis obviously begins with light absorption by antennae pigments and the primary photochemistry in the reaction centre. It is necessary to understand the occurrence of several hundred antennae pigments for every given reaction centre. The chl a molecule has a cross-sectional area of 0.67 $Å^2$, which at peak light intensity is 2000 μ mol photons $m^{-2} s^{-1}$. The rate of light absorption for chlorophyll is 8 photons/second. Obviously, this is insufficient for driving the overall photosynthetic reaction (Owens, 1996). Plants have, therefore, evolved systems for overcoming this inadequacy by associating several hundred antennae pigments.

1.3. STRUCTURE AND ROLE OF ANTENNAE

In higher plants and algae, antennae systems are unique for PS I and PS II which differ in pigment and protein composition.

Both the photosystems possess antennae which can be conveniently divided into core (or proximal) and peripheral antenna complexes (Paulsen, 1995, Bricker and Frankel, 2002, Psylinakis et al., 2002). A schematic view for the overall organization of the components of the antenna system is presented in Figure 1.2. Although the core antennae complex comprise pigments in a fixed stoichometry, the peripheral antennae complex have variable pigments (Bassi et al., 1990). Apparently, the size and composition of peripheral antennae is very much under the control of environmental regimes. LHC I and LHC II are the peripheral antennae of PS I and PS II, respectively, and are composed of chlorophyll a/b binding proteins. Additionally, PS II contains three minor peripheral complexes, CP 29, CP 26 and CP 24, which connect the main LHC II and PS II core (Bassi and Dainese, 1992).

In brief, the light harvesting system of PS II can be viewed as composed of two clearly distinguishable types of pigment protein

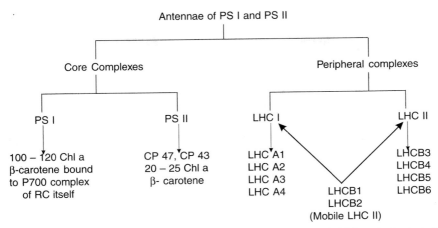

Fig. 1.2. The organization of light harvesting antennae system in higher plants. In both PS I and PS II, the antennae system is composed of core and the peripheral complexes. The individual protein subunits comprising each of these components are listed, including the mobile LHC II which serves both the photosystems.

complexes. The core antennae linked to reaction centre consists of two chlorophyll-binding proteins, CP 47 and CP 43 exclusively with chlorophyll a molecules (Bricker and Frankel, 2002). Also, CP 47 and CP 43 are known to interact with water oxidation proteins. They are apparently required for obtaining high rates of water oxidation (Bricker and Ghanotakis, 1996; Bricker and Frankel, 2002). These proteins form the binding sites for the extrinsic proteins of PS II, a conclusion well supported by the work of Zouni et al. (2001). The other antennae, which are bound to the PS II core forming the light-harvesting complexes of LHC II with polypeptides ranging 20 to 30 kDa, bind both chlorophyll a/b and xanthophylls (Jansson, 1994; Paulsen; 1995; Horton et al., 1996). They are called Chlorophyll a/b binding CAB polypeptides found in thylakoid membranes of chloroplasts. There exist ten distinct types of CAB polypeptides encoded by nuclear genes, collectively designated as Lhc genes. The chlorophyll pigments a and b assume specific orientation by association with proteins known as CAB polypeptides. These polypeptides bind roughly 50 per cent of Chl a, all of Chl b and most of xanthophylls, lutein, neoxanthin, violaxanthin, antheraxanthin and zeaxanthin.

Ten distinct LHC proteins are recognized in higher plants (Jansson, 1999) which constitute subunits of both PS I and PS II. Of these ten, four of CAB proteins — LHC A1, LHC A2, LHC A3 and LHC A4 are associated with PS I, while LHC B1-LHC B6, constitute the antenna of PS II. Thus, four proteins, LHC B3, LHC B4, LHC B5 and LHC B6 are regarded to be exclusively associated with PS II. The two proteins LHC B1 and LHC B2 are regarded to be serving either of the photosystems, although they are known to be constituents of LHC II. The CAB proteins play a key role in light harvesting. Besides this role, they are also involved in acclimation to varying light regimes (Pichersky and Jansson, 1996; Jansson, 1999). Other essential functions of LHC include a photoprotection through quenching of triplet chlorophyll and in mitigating photoinhibition by dissipation of excess excitation energy (Formaggio et al., 2001). The corresponding genes are named as Lhc genes (Jansson et al., 1992; Jansson, 1999). The LHC proteins are composed of three membrane spanning helices. Helix one and three are homologous. They also share a generic LHC motif. This is a sequence of 22 amino acids, of which three — glutamic acid, arginine and glycine — are invariable components. The helices form the core of the complex and they are held together by Arg – Gln. salt bridges (Kuhlbrandt et al., 1994). Most of the pigment molecules and carotenoids are bound to these homologous regions.

    The main component of LHC II, serving photosystem II, is referred to as LHC IIb. This is the most abundant complex and binds about 45 per cent of the total chlorophyll. The remaining LHC II components bind 4 per cent of chlorophyll each. In addition, the above proteins of LHC II contain xanthophylls, lutein, neoxanthin and violaxanthin in varied amounts (Bassi et al., 1993; Morishige and Dreyfuss 1998; Croce et al., 1999; Pineau et al., 2001). The LHC II b has the trimeric structure in all plant species so far examined and is composed of polypeptides of sizes 28, 27 and 25 kDa, which are encoded by the genes Lhc b1, Lhc b2 and Lhc b3, respectively (Paulsen, 1995; Horton et al., 1996). Yang et al., (2003) have recently studied the folding behaviour of the apoprotein Lhc b1 in the formation of functioned LHC IIb in vitro. The study has shown that the initiation of α-helix formation is not a rate-limiting step in LHC IIb synthesis, although an α-helical segment is a prerequisite for other proteins (Yang et al., 2003).

    The Lhcb1 by itself is a homotrimer. Lhcb1 and Lhcb2 together form the heterotrimer, both of whom also function as an antenna for either of the photosystems. The trimer containing Lhc b_3 formed from 2 Lhc b1 plus one Lhc b3 serve exclusively PS II (Jansson, 1999). Lhc b4, Lhc b5 and Lhc b6 proteins (called CP 29, CP 26 and CP 24, respectively) are monomeric and are present only in one copy per PS II unit (Jansson, 1999). Therefore, Lhc proteins of PS II include Lhcb 1-3 gene products form the heterotrimeric complexes (LHC II) which are present peripherally in the overall supercomplex of PS II – LHC II, while the proteins Lhc b 4-6 (CP 29, CP 26 and CP 24) are monomeric and situated in between the trimeric complex and PS II core (Dominici et al., 2002). The PS II major antenna is organized into a mobile complex formed by Lhcb1 homotrimers and heterotrimers of Lhcb1 and Lhcb2 together as well as an immobile

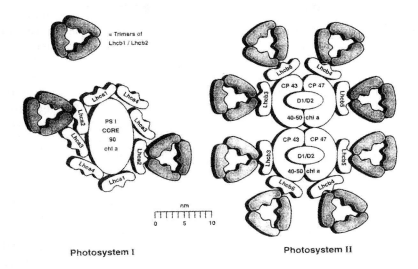

Fig. 1.3. The organization pattern for the Light Harvesting Antennae of plants. The pattern is actually dependent on the stoichiometries of the individual protein components. The model in this figure shows a single layer of Lhca polypeptides around the core antennae of PS I. The outer antennae, consisting of Lhcb 1/2 trimers, is connected to Lhca2 polypeptide. The antennae of PS II (shown in dimer form) consists of a core CP 43 and CP 47 subunits sorrounded by an inner antennae of Lhcb 3,4,5 and 6 and an outer antennae of Lhcb 1/2 trimers. (Figure reprinted with permission by Kluwer Academic Publishers from D.J. Simpson, J. Knoetzel 1996 Light Harvesting complexes of plants and algae. In : Advances in photosynthesis Vol. 4 pp.493-506. Copy right 1996 Kluwer Academic Publishers.

complex of Lhcb1 and Lbcb3(2:1). Figure 1.3 represents the topological organization of the proteins in the antennae system, based on their stoichometrics as per Knotzel et al., (1992) for PS I and Jansson (1994) for PS II, as shown in a model by Simpson and Knoetzel (1996). The immobile complex is structurally associated with the minor antenna protein (CP 29, CP 26 and CP 24) that surrounds the core complex. The architecture of LHC II involves the minor antenna linking the core of PS II and major antenna. The periphery of PS II is occupied by major antenna (Bassi and Dainese, 1992, Huber et al., 2001). Several isoforms of the Lhcb proteins have been found, particularly in Lhcb1 (Huber et al. 2001). Lhcb3 existed in 2 isoforms in *Petunia* and tomato, while Lhcb6 had the largest number of isoforms in *Petunia*, tobacco, tomato and rice. The other proteins found in only one form are Lhcb2, Lhcb4 and Lhcb5. The existence in multimeric forms of Lhcb1 and, to a lesser extent, the Lhcb3 and Lhcb6 may indicate a role in the adaptation of plants to environmental stresses (Huber et al., 2001).

The work of Kuhlbrandt et al., (1994) is the most crucial development in the area of photosynthetic research, related to chlorophyll carotenoid proteins, which have resulted in our present understanding of the atomic structure of LHC II. Unravelling the structural configuration of LHC II is essential in order to appreciate its role in light harvesting but also providing a generalized model of folding for all the Chl a/b proteins (Green and Durnford, 1996). The crystal structure of Lhc II b, as determined by Kuhlbrandt et al., (1994) and the atomic model of LHC II proposed therein indicated that each protein binds 12 chlorophyll molecules (seven Chl a and five chlorophyll b molecules). Additionally, two structurally-important lutein molecules form an internal cross base (Kuhbrandt et al., 1994). The atomic structure of a monomer of LHC II is shown in Figure I.4. Other carotenoids, neoxanthin and violaxanthin, are probably at the perifery of the trimer (Flachman, 1997).

Besides, the major complexes of LHC II b, the LHC of PS II contain three minor complexes CP 29 (LHC II a), CP 26 (LHC II c), CP 24 (LHC II d). All of these are monomeric and bind only 5 per cent of PS II chlorophyll. Three transmembrane spaning α - helices (A, B and C) are observed in the LHC II b crystal structure at 3.4 Å resolution. Thus, LHC II polypeptide folds in the three–membrane spanning helices connected by hydrophilic loops with an additional short amphiphilic helix located at the lumenal surface of the thylakoid. Further, the model proposed by Kuhlbrandt et al., (1994) has suggested an arrangement of chlorophylls perpendicular to the plane of the membrane in layers close to stromal and lumenal surfaces. Two lutein molecules are embedded with in the LHC II b complex (Tremoliers et al., 1994; Bassi et al., 1992; Jansson 1994) suggested that the overall LHC II system contains five units of LHC II b and three to four minor complexes, constituting an overall oligomeric intrathylakoid antenna component (Horton et al., 1996).

Carotenoids are present in all the PS I and PS II pigment proteins. β-carotene is bound mainly to the core complex of PS II, while xanthophylls are present on the antenna system (Yamamoto and Bassi, 1996). Chloroplast-encoded subunits of PS II are chlorophyll a, β carotene complexes, while nuclear-encoded subunits of antennae system are chlorophyll a/b and xanthophyll complexes. Each pigment protein of antennae system has a particular carotenoid composition in addition to a common lutein component (Yamamoto and Bassi, 1996).

There has been no uniform nomenclature (Paulsen, 1995) for the description of chlorophyll a/b apoproteins. For instance, the nomenclatures evolved by Bassi et al., (1990), Thornber et al., (1994) and Jansson et al., (1992) exist in literature and are used from different laboratories. A comparative account of the different nomenclatures and their equivalents has been reviewed by Paulsen (1995). In the present context, no single uniform nomenclature has been used throughout the text for the apoproteins. At several places, the equivalent term of common practice has also been indicated.

A specific nomenclature for the genes encoding CAB polypeptides has emerged (Jansson et al., 1992). This particular nomenclature leads to a division of the genes encoding CAB proteins into two groups as follows:

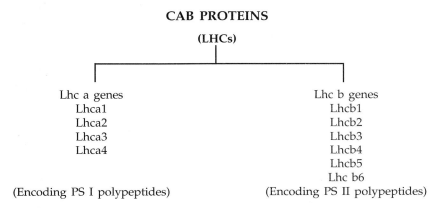

Lhca1 to 4 genes encode the subunits of LHC I. Lhc a, and Lhc a4 together term the complex referred to as LHC I – 730 (Chl a/b ratio 2.3), while Lhca2 and Lhca3 constitute the complex LHC I – 680 (Chla/b ratio = 1.4 Jansson 1994, Funk 2001). Lhc b genes encode the subunits of LHC II, which includes the major LHC IIb (LHC II) complex along with three minor subunits (Jansson et al., 1992, Jansson 1999).

CAB proteins have the ability to adjust the functional antenna size in response to light conditions. It has been shown recently (Teramoto et al., 2002) that the expression of genes for PS II is coordinately repressed under high light conditions in *Chlamydomonas reinhardtii*. However, it was also found that high light repression of Lhc b genes occurs in mutants deficient in PS II or in both PS I and PS II (Teramoto et al., 2002). It was further included that two different mechanism of Lhc gene repression might exist (Teramoto et al., 2002).

LHC B1 : This is the most abundant CAB protein and has been used as a model system with its gene Lhcb1. Having been identified in eighteen species of plants (Jansson et al., 1992), this also constitutes LHC II. LHC B2 is the another component of the LHC II. Functionally, B1 and B2 form the mobile LHC II and participate in reversible phosphorylation. It is also considered to be involved in the long-term acclimation to different light regimes. LHC B3 is both of a smaller size and less abundant than B1 and B2. This is isolated from tomato, pea, barley and *Brassica*. LHC B4, the largest of CAB proteins, is considered to possess common characteristics with LHC A1 and LHC B5. LHC B4, B5, B6 and perhaps B3 are sometimes referred to as inner LHC II which are otherwise known as minor complexes (the part of LHC II, not disconnected from PS II, after phosphorylation of mobile LHC II, which means LHC b1 and Lhc b2) (Jansson 1994, Pichersky and Jansson, 1996). LHC B5 is more likely CP 26. LHC B6 is otherwise known as CP 24 (LHC II D). This is the smallest of all known LHC B proteins. Except for the Lhcb1 gene, the other nine genes have been investigated in tobacco and scotspine and, to a lesser extent, in *Arabidopsis*. The Lhc genes are expressed in photosynthetic tissue. The studies have shown that all the ten types of CAB proteins are always found and that Lhc b1 is the most abundant.

Structure of Lhcb1 gene product has been resolved at 3.4 Å resolution by Kuhlbrandt et al. (1994). The protein has three membrane-spanning regions, which are in the α - helix configuration. The structure has been elucidated as follows : 80 per cent of polypeptides, 12 chls and two luteins are fitted and 8 or 9 chl ligands have been identified (Green and Durnford, 1996, Simpson and Knoetzel 1996). The pigment content estimated using biochemical data appears to vary from the above, which indicates 8 chl a, 6 Chl b, 2 luteins, 1 neoxanthin, and 1 zeaxanthin, bound to each polypeptide (Pichersky and Jansson, 1996).

LHC A1 together with LHC A4 are the shortest of LHC proteins with 200 residues. Their functions are yet to be established. The sequences of Lhc a1 gene have been characterized from five plant species tomato, *Arabidopsis*, tobacco, scotspine and barley, while that of LHC A2 (LHC I-680) from four plant species; tomato, *Petunia*, scotspine and barley. LHC A3 is the largest of the LHC A group of polypeptides with a molecular weight 24 to 25 KDa. It is present in the LHC I – 680 pigment protein complex. The gene sequence of this Lhca3 is characterized in tomato, peas, scotspine and potato. LHC A4 is associated with LHC A1 in the LHC I 730 complex (Pichersky and Jansson 1996; Morishige and Dreyfuss 1998). LHC A polypeptides are stable even in the absence of Chl b, as evidenced by their equal abundance in barley mutant chlorina f2, which is less in Chl b (Krol et al., 1995).

In addition to the ten LHC proteins described, there are additional complexes in the light-harvesting antennae of PS II. In this respect, CP 22, the PsbS gene product is an important novel component of PS II (Jansson, 1999; Morishige and Dreyfuss, 1998; Li et al., 2000, Funk 2001). The PsbS protein is an intrinsic component of PS II and differs from the other CAB proteins in view of its possessing four transmembrane helices instead of three in all CAB proteins. The PsbS protein is apparently located close to the PS II core complex, probably connecting the minor light harvesting proteins and the core antenna (Funk 2001). It is considered to bind the pigments Chl a and Chl b but does not take part in light harvesting although it does particiate in energy dissipation (Li et al., 2000; Funk

2001). However, the results of Dominici et al., (2002) have shown that PsbS does not bind pigments or if at all it does so, then the binding is different from other LHC proteins. This finding of Dominici (2002) is also consistent with the earlier result of Funk et al., (1995) that PsbS is stable in etiolated leaves, which obviously lack chlorophyll. Among other novel proteins of PS II, Ruf et al., (2000) have described the existence of a chloroplast-encoded protein coded by a Ycf 9 gene as a structural component of LHC of PS II. This protein is of significance for the integration of the LHC protein CP 26, into antennae of PS II. Recently, Ruban et al., (2003) have investigated the macroorganization of PS II in plants lacking main light harvesting complex (LHC II). They have used antisense *Arabidopsis* plants which lack LHC II trimers and studied the formation of the PS II supercomplexes. It was observed that in the absence of the main LHC, another protein is substituted for the assembly of PS II. In fact, CP 26 (Lhcb5 gene) has replaced the LHC II (Lhc b1, and Lhcb2). This replacement, however, has not produced the activity of state transitions inspite of the replacement of the main complex. Therefore, the macrostructure can be retained by replacement of missing antennae, as evidenced by the work of Ruban et al. (2003).

The size of Chl antenna is usually determined by different approaches, including spectrophotometric, optical cross-section and biochemical method (Melis, 1996). A minimal size of chlorophyll antennae in PS II is estimated to be 37 Chl a molecules located in PS II core. The core would comprise the heterodimer D1, D2, cyt b 559, low molecular weight proteins and the inner antenna CP 43 and CP 47. The LHC II, the light-harvesting antennae when complexed with the core, the overall size of the antennae system would increase. The actual sizes in both PS I and PS II, respectively, in a variety of higher plants and algae are 210 and 230 for spinach 220 and 280 for pea 240 and 530 for *Chlamydomonas*. (Melis, 1996) The size of the antennae is very much dependent on the level of irradiance (Melis, 1991). It is understood that low growth light would lead to larger antenna size for both the photosystems, while higher light intensities are associated with relatively smaller chlorophyll antennae size. Thus, there is a considerable flexibility of the size of the antennae system depending on the light environment.

It is now well understood that LHC have multiple functions in the overall photosynthetic process. Recently, Formaggio et al., (2001) have made an extensive study of the functional architecture of major LHC II from maize. The LHC proteins have at least fourfold functional characterstics. These include light harvesting (Havaux et al., 1998) and energy transfer, photoprotection by quenching ^{3}Chl, regulation of light harvesting by quenching ^{1}Chl, and a very specific role in providing stability of protein folding and in the prevention of the synthesis of unprotected chlorophyll protein complexes (Demmig – Adams, 1990, Niyogi, 1999, Formaggio et al., 2001; Hirschberg, 2001). Photoprotection and related functions are discussed in greater detail in Chapter 3. Formaggio et al., (2001) have modified the carotene and chlorophyll binding sites and obtained mutants related to the three structural domains. They have suggested that NPQ is regulated by an allosteric modification of protein structure achieved through xanthophyll exchange at different sites.

Fluctuations in the quality of light might occur, leading to absorption of light energy in an unbalanced manner by preferential harvesting of light through either of the two photosystems PS II or PS I. Such a situation creates an unbalanced electron flow. This rapid process is known as state transition and is correlated with phosphorylation and migration of LHCP (LHC II) between PS II and PS I. State transitions are a mechanism for the response of the plant to sudden fluctuations in light quality (Melis, 1996; Veeranjaneyulu et al., 1998; Pfannschmidt et al., 1999). Recently, Snyders and Kohorn (2001) have studied in detail the protein kinases involved for the phosphorylation of LHCP and the regulation of state transitions. They have shown a requirement of TAKs (Thylakoid Associated Kinases) in state transition due to fact that the loss of state transitions was associated with reductions in TAK and loss of LHCP phosphorylation.

The function of PS II reaction centre is dependent on the efficiency of transfer of energy from the antennae to the reaction centre. There are two views regarding this; one being a Reversible Radical Pair Model (RRP) which involves a rapid equilibration of excitation energy between the antennae and the reaction centre (Barter et al., 2001). The second is known as Energy Transfer to the trap limited model (Vasil'ev et al., 2001; Diner and Rappaport, 2002).

1.4. COMPOSITION OF PHOTOSYSTEMS

The antenna system comprising the proteins of the light harvesting complexes (LHC I and LHC II) constitutes part of the overall supramolecular complexes PS I and PS II with the corresponding reaction centres of PS I and PS II.

Oxygenic photosynthesis is characterized by the physical structures of two photosystems — photosystem II (PS II) and photosystem I (PS I) — driving electron transport from water to $NADP^+$. The holocomplex of each photosystem is composed of a core complex performing light energy transduction surrounded by LHC responsible for light harvesting antennae function. Molecular mechanisms of PS II could obviously be understood, provided a detailed knowledge is available of the three dimensional structures, particularly of the reaction centre which brings about the unique reaction of water oxidation and oxygen evolution (Nield et al., 2000).

The knowledge of molecular structure of the membrane protein complex of PS II and PS I is essential for a precise understanding of its role. For quite some time before the advent of the modern technological advances, models for subunit organization of supramolecular structure of the photosystem complex have been gained through biochemical fractionation and chemical cross-linking studies. However, relatively recently, a knowledge of the structural details based on electron crystallographic studies of photosystem II in spinach have been gained sequentially, beginning with low resolution analysis (Barber et al., 1997; Hankamer et al., 1997a), followed by a relatively high resolution of 8 Å (Rhee, 1998) and 6 Å (Rhee 2001, Hankamer et al., 2001). Currently, x-ray crystallographic studies have led to 3D structural models of PS II core from cyanobacterium, *Synechococcus* at 3.8 Å (Zouni et al., 2001) and, subsequently, by Kamiya and Shen (2003) in *Thermosynechococcus* at 3.7 Å. The higher plant PS II is yet to be investigated at its atomic structural level (Rhee 2001). Also, x-ray crystallographic analysis is still not sufficient to denote the orientation of amino acid residues. Though resolution is still low for the determination of side chains of amino-acid residues, most

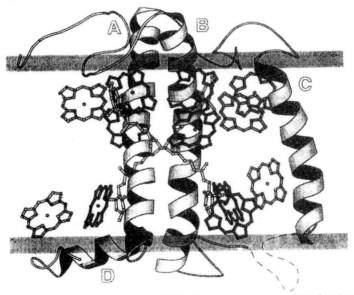

Fig. 1.4. Atomic model of a monomer of LHC II. The boundaries represent the stromal (top of the figure) and lumenal sides of the thylakoid membrane. The ribbon structures are the α - helices labelled ABC which span the membrane and the extension helix D. Two letein molecules brace the helices A and B forming a cross between them in the center of the complex. The location of chlorophyll molecules (shown as tetrapyrrole rings) is depicted. (Reprinted by permission from Nature, W. Kuhlbrandt et al., 1994 Atomic Model of Plant Light — Harvesting Complex. Nature Vol. 367 pp. 614-621. Copyright 1994 Macmillan Publishers Ltd.)

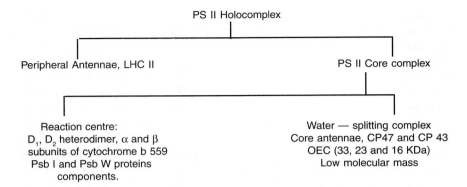

Fig. 1.5. Schematic representation of the components of PS II holocomplex in plants. The PS II supercomplex is composed of a core complex and a peripheral antenna, LHC II. The subunits in each of these components are listed.

chromophores, α-helices, β-sheets and Mn-cluster have been located (Diner and Rappaport, 2002). It is further suggested that the use of random mutagenesis is going to supplement the x-ray-diffraction crystallographic studies (Yamasato et al., 2002), while site-directed mutagenesis was applied earlier (Diner et al., 2001) to analyze the structure of PS II.

PS I in oxygenic systems at 4 Å structure has become clearer relatively recently (Krauss et al., 1996). Also, Jordan et al., (2001) have determined the 3-dimensional structure of photosystem I in the cyanobacterium, *Synechococcus elongatus* at 2.5 Å resolution (Kuhlbrandt, 2001). This study has provided the atomic detail of protein subunits and cofactors of photosystem I.

From sequence analysis, it has become clear that purple bacterial RC is similar to that of PS II.

The structural details of PS II could be obtained by different physical methodological analysis. Using electron microscopy, the size and shape of PS II could be established at a comparatively lower resolution of 40 – 50 Å. At higher resolutions, upto 15 Å, two additional approaches are adopted. These methods include single particle image and analysis of two-dimensional crystals (Hankamer 1997a; Hankamer et al., 1997b). It is considered that the former method, the single partical image, results in providing information relating to oligomeric status and subunit positioning of PS II. The method of two-and three-dimensional crystallography offers material for structure determination at atomic resolution (Mo, 1995).

1.5. PHOTOSYSTEM II : COMPOSITION AND STRUCTURE

Till recently, the conceptual models describing the structure of photosystem

II, principally derived by Electron Microscopy of intact PS II at 15 – 30 Å resolution have been available (Nield et al., 2000). Nield et al., (2000) have determined the three-dimensional structures of PS II complexes from *Chlamydomonas* and *Synechococcus* at 30 Å resolution. The particle isolated from *Chlamydomonas* was found to be similar to LHC II – PS II super complex from spinach. Further, the *Chlamydomonas* PS II complex was dimeric and also similar in several structural features of OEC subunits to that of spinach.

There are altogether over 25 different proteins in PS II complex (Barber et al., 1997; Hankammer et al., 1997; Barber and Kuhlbrandt, 1999). The reaction centre is composed of D1 and D2 proteins which bind all the cofactors involved in the primary and secondary electron transport and are located in the centre of the complex. The architecture and the subunit composition of PS II complex (Hankamer et al., 1997) is presented in Figure 1.6. Further, the manganese cluster responsible for water

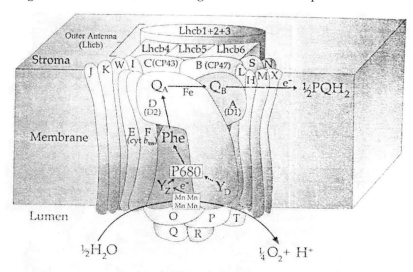

Fig. 1.6. The subunit composition of Photosystem II and the path of electron transport. The relative positions of polypeptide components of PS II encoded by the genes PsbA to PsbX in the organization of the overall photosystem II is presented. The notation of the subunits is based on the genes (A-X) of which they are the products. The core complex is surrounded by minor antennae proteins Lhcb4, Lhcb5 and Lhcb6. The outermost antennae layer consists of Lhcb1, 2 and 3. The electron transport components are located on the D1/D2 heterodimer of reaction centre. Manganese cluster and the oxygen evolving complex are located on the lumenal side of the membrane. (Figure reprinted with permission by Annual Reviews from B. Hankamer et al., (1997) Structure and membrane organization of photosystem II in green plants. Annual Review of Plant Physiology and Molecular Biology volume 48 pp.648-671. Copyright 1997 by Annual Reviews www.annualreviews.org).

oxidation is regarded to be ligated to the D1 on the lumenal side. The other subunits of PS II surround the heterodimer D1 and D2. Among the subunits, CP 47 and CP 43, which transfer excitation energy to the reaction centre, are essential (Debus, 2000).

A model for the structure and organization of transmembrane helices of CP 47 and CP 43 (Six transmembrane helices) and other subunits has been presented through studies by electron crystallography (Rhee et al., 1998; Hankamer et al., 1999). Besides the inner antennae of CP 47 and CP 43, the higher plants and green algae additionally possess the peripheral light harvesting system of Chl a and Chl b. Hankamer et al., (2001) have described the 3D structure of PS II core dimer of spinach through electron crystallography. The resolution through electron microscopy has allowed the assignment of the transmembrane helices. There are 34 transmembrane helices for each unit of the dimer. Twenty-two helices are assigned to the subunits D1, D2, CP 47 and CP 43. Twelve helices are assigned to low molecular weight proteins which have single transmembrane helices. They have also made a comparision of the subunit organization of the higher plant PS II core dimmer reported by them with cyanobacterial PS II core, as determined by x-ray crystallography (Zouni et al., 2001). Zouni et al., (2001) have determined the crystal structure of PS II from *Synechococcus*, at 3.8 Å resolution, showing a three-dimensional structure of water oxidizing PS II complex. These authors (Zouni et al., 2001) have described the spatial distribution of the different subunits and the cofactors in PS II and also with a detailed analysis of the structure of manganese cluster relating to its location, shape and size. They have isolated PS II as homodimers and used them for growing 3 D crystals for x-ray analysis. PS II was composed of at least 17 subunits (Zouni et al., 2001). Fourteen subunits comprise the reacton centre proteins, D1 and D2, inner antennae subunits CP 43 and CP 47; α and β subunits of cytochrome b-559 and low molecular weight components, PsbH, PsbI, PsbJ, PsbK, PsbL, PsbM, PsbN and PsbX. Also, the extrinsic units PsbV, PsbU and PsbO (33 kDa proteins, manganese stabilizing) are on the lumenal side of PS II (Barry et al., 1994). PS II occurs as a homodimer with the longest dimensions 190 Å x 100 Å. It is 40 Å thick, extending from the stromal side by 10 Å. The lumenal side of each monomer protrudes by 55 Å. The monomers are related by local C2 rotation axis. Based on the arrangement of the α - helices of D1 and D2, it was concluded that there was a resemblence of this structure to that of purple bacterial RC subunits, L and M. Accordingly, this finding supports the concept of common ancestry for all the photosynthetic reaction centres. The location of inner antennae subunits CP 47 and CP 43 was found to be on the two sides of the complex. The study conducted by Zouni et al., (2001) has established the fact that the assignment of subunits D1, D2 and CP 47 corresponds to the model provided by Rhee et al., (1998) and Rhee (2001) at 8 Å resolution of PS II in higher plants.

Diner and Rappaport (2002) have reviewed the recent work in connection with 2 D and 3 D Electron and x–ray crystallography of the PS II core complex. The authors have brought out the contemporary research in the areas of structure and energetics of photosystem II. The role of cofactors of PS II, particularly tyrosine residues, is currently an important area of research and several workers have drawn their attention to these aspects (Faller et al., 2001; Ananyev et al., 2002).

As already stated, the complete atomic resolution for PS II is lacking (Rhee 2001). However, biochemical fractionation and chemical cross linking have given rise to information on the supramolecular organization of the complex, leading to the models for protein subunit organisation. The overall PS II is made up of the LHC II and the PS II core complex (Bricker and Ghanotakis 1996; Tsiotis et al., 1999; Psylinakis et al., 2002; Barber, 2002). Figure 1.5 is a summarized diagrammatic representation of the composition of the PS II holocomplex in higher plants. PS II core complex comprises the reaction centre (RC), two chlorophyll, a binding core antenna CP 47 and CP 43, water-oxidizing complex (33,23,16 kDa) and several low molecular mass proteins (Ghanotakis and Yocum 1986).

The Reaction Centre is composed of the heterodimer, the D1 and D2 α and β subunits of Cyt b 559, PsbW and PsbI proteins (Nanba and Satoh 1987; Fotinou and Ghanotakis 1990; Psylinakis et al., 2002). The various components of the holocomplex are depicted diagrammatically.

The structure of PS II from both higher plants and cyanobacteria through electron microscopy and single particle imaging was studied. The O_2 evolving core complex has a molecular mass of 450kDa and dimensions are 17.2 x 9.7 nm, and showed a twofold symmetry, indicating a dimeric organization (Boekema et al. 1995; He and Malkin 1998).

The intact PS II complex with LHC II has a mass of 700 kDa and it is a dimer with dimensins 26.8 x 12.3 nm (He and Malkin 1998). Therefore, two LHC II trimers are present in each of the PS II complex (dimeric). Each of the trimers is linked to PS II core complex by LHC proteins, CP 24, CP 26 and CP 29. Accordingly, PS II might exist in vivo as a dimer (He and Malkin 1998).

1.5A. Reaction Centre of Photosystem II

The electron transfer, primary charge separation, water oxidation and photo-oxidation reactions take place in the RC of photosystem II. PS II RC contains six Chl molecules, two pheophytins (Pheo), two β carotenes and two plastoquinones (Q_A and Q_B), bound to D2 and D1, respectively (Yoder et al., 2002). The D1 and D2 polypeptides occurring pseudosymmetrically about C_2 axis show the arrangement of the ten transmembrane α - helices. D1 and D2 have five transmembrane helices and two histidine residues D1 – H 118, D2 – H 118 are conserved in PS II. A monomeric chlorophyll ChlZ is electron donor to P 680 under certain circumstances. Probably, D1-H 118 is the ligand to ChlZ (He and Malkin, 1998).

By definition, PS II RC is constituted of D1 and D2 heterodimer with α (9 kDa) and β (4 kDa) polypetides of cytochrome b 559 and the psbI and psb W gene products (Tracewell et al., 2001). CP 43 and CP 47 (43 and 47 kDa) are located on the opposite sides of the reaction centre. Clear evidence has been presented by Barber et al., (1999) to revise the earlier notion (Rogner et al., 1996) of the location of the CP 47 and CP 43. Barber et al., (1999) have assigned the subunit position in relation to CP 43 and CP 47 establishing that CP 43 is located adjacent to D1 and that CP 47 to D2. This is also confirmed by the fact that during turnover, D1 protein CP 43 is released. Therefore, proteins with a molecular weight of 47 and 43 kDa (CP 47 and CP 43) are bound to RC on the opposite sides. CP 47 and CP 43 also form binding sites for the extrinsic proteins of PS II (Bricker and Frankel, 2002). This reasoning is based on the x-ray study made Zouni et al. (2001). Both proteins are involved directly in the O_2 evolving process. They interact with the three extrinsic water oxidation proteins, 33,23 and 17 kDa in size. The role of these proteins in water oxidation is in addition to their function as proximal antennae proteins of the PS II (Bricker and Frankel 2002). Recently, Henmi et al., (2003) have investigated the interaction of D1 protein, CP 43 and OEC 33 at the lumenal side of PS II. The results are also consistent with the assignment of the subunits in the 3D structure based on x-ray diffraction analysis and crystal structure (Zouni et al., 2001; Kamiya and Shen, 2003). Two Chl a molecules are located at the centre of D1, D2. Formation of exited P680* leads to electron transfer from P680 to pheophytin, generating the charge separated state $P680^+$ and Pheo. The structural unit consisting of D1, D2 cyt b 559, PsbI, PsbW gene products of RC core is responsible for the charge separation.

Higher resolution electron crystallography studies have revealed the new structural data for the location of major subunits of PS II core and their transmembrane helices (Barber et al., 1999).

The authors (Barber et al., 1999) have proposed an improved model of PS II. They have isolated and characterized a large dimeric complex from spinach, which contains majority of subunits and also the light harvesting Chl a/b complex (LHC II) as well as PS II core RC proteins. It is termed as the LHC II – PS II super complex. The model suggests the possible location of two inner antennae CP 47, CP 43 relative to the RC proteins, D1 and D2.

I.5B. Core Complex of Photosystem II

The LHC II – PS II supracomplex contains a dimeric PS II core, flanked by two sets of Chl a and Chl b binding proteins. The subunits of PS II core are usually divided into two functional groups :

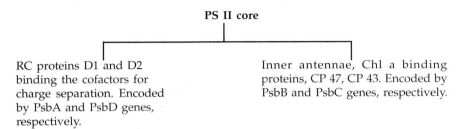

RC proteins D1 and D2 binding the cofactors for charge separation. Encoded by PsbA and PsbD genes, respectively.

Inner antennae, Chl a binding proteins, CP 47, CP 43. Encoded by PsbB and PsbC genes, respectively.

According to this model (Barber et al., 1999), CP 47 and CP 43 must be located on either side of the D1, D2 heterodimer. D1 transmembrane helices in the D1, D2 heterodimer are flanked by the helices of CP 47 and CP 43. In the model, the D1 and D2 proteins are adjacent to CP 43 and CP 47, respectively (Barber et al., 1999).

Two tyrosine radicals and tetramanganese cluster are on the oxidizing side of PS II. The tyrosine radicals are identified as D1 tyr 161 (YZ) and D2 tyr 160 (YD), respectively (He and Malkin, 1998; Faller et al., 2001). The structure of Mn cluster has not been resolved. The arrangement of three transmembrane helices and one lumenal surface helix of LHC II trimer is based on the published 3.4 Å sturucture. The positioning of minor LHC II proteins, CP 29, CP 26 relative to CP 47 and CP 43 is based on recent findings by Harrer et al. (1998). The model is consistent with the cross-linking data and serves as a basis for further work on molecular processes involved in water spliting and oxygen evolution.

1.6. PHOTOSYSTEM I – STRUCTURE AND ORGANIZATION

Photosystem I is a membrane pigment protein complex containing several subunits. In higher plants, it is composed of 18 subunits, of which at least 14 subunits (Scheller et al., 2001; Chitnis, 2001) have been found in PS I core, while in cyanobacteria, only 11 polypeptides have been identified. These subunits are denoted as PsaA to PsaN or PS I-A to PS I-N. The structure of PS I, the function of its subunits and related aspects have been reviewed in detail (Golbeck, 1992; He and Malkin, 1998; Scheller et al., 2001; Chitnis, 2001). A 3-dimensional structure of cyanobacterial PS I at 2.5 Å resolution has been described by Jordan et al., (2001). The work of Jordan et al., (2001) determining the x-ray crystal structure of PS I, has provided the considerable advancement over the previous knowledge of the structure. This study is expected to answer several areas of PS I function. Further, in recent years, work relating to PS I subunit organization (has been carried out by several anthors Ihalainen et al., 2002; Varotto et al., 2002; and Jensen et al., 2002).

All PS I complexes have two high molecular mass subunits of 83 kDa coded by PsaA and PsaB (PS I-A) (PS I-B) present in single copies and they form a heterodimer binding P 700, Ao, A_1, Fx as well as 120 Chl a. The core proteins contain 11 transmembrane helices. The terminal electron acceptors in PS I, F_A and F_B are bound by a low molecular mass subunit of approximately 9 kDa, known as PsaC (PS I-C). Additional binding subunits of PS I are less well defined. PsaD and PsaF have been implicated in the interaction of PS I with the electron transfer partners Fd and Fc, respectively (He and Malkin, 1998). Other subunits PsaG, PsaH, PsaI, PsaJ, PsaK, PsaM and PsaN have no known functions. Cyanobacteria do not contain PsaG and PsaH. In cyanobacteria, PS I contains 125 Chl a+b per P 700, while in higher plants, PS I complex has 225 Chl a+b molecules per P 700 (He and Malkin, 1998).

I.6A. Reaction Centre of Photosystem I

The reaction center of the photosystem is a light-driven plastocyain : ferredoxin oxido reductase. It is a pigment protein complex in higher plants, green algae and cyanobacteria. The primary electron donor is P_{700} and primary electron acceptors are Ao, an intermediate quinone acceptor A_1 and three iron–suphur centres, Fx, F_B, F_A. The charge separation is begun by photon absorption by one of the two hundred antennaa chlorophyll molecules associated with each RC. Photochemical charge separation is brought about between P 700 and Ao. The electron on Ao$^-$ is shuttled through A_1 and Fx to the terminal acceptors F_B/F_A. The charge separation permits the occurrence of electron transfer between P 700$^+$ and plastocyanin, and between F_{A-}/F_{B-} and ferredoxin. The reduced ferredoxin on the structural side and oxidized plastocyanin on the lumenal side are crucial for charge separation (Golbeck, 1992).

PsaA (PS I-A) and PsaB (PS I-B) proteins forming the RC heterodimer with molecular masses of 82-83 kDa contain hundred antennaa chlorophyll molecules. Phylloquinone (Vitamin K_1) may be the A_1 intermediate electron acceptor; it may also reside on the PsaA/PsaB heterodimer. The iron-sulphur centre Fx bridges the PsaA and PsaB heterodimer. Finally, the F_A/F_B acceptors exist as two independent 4 Fe-4S clusters located on a 8.9 kDa polypeptide, PsaC. The higher plant PS I holocomplex contains 10 additional polypeptides labelled PsaD through PsaL and 4 and 5 light-harvesting chlorophyll proteins (LHC I). The functions of many of the low mass proteins are unknown. Cyanobacterial PS I complex contains 10 additional polypeptides PsaC to PsaF and PsaI through PsaN and lacks the entire LHC I. A detailed analysis of the structure of a cyanobacterial PS II complex has been conducted by Jordan et al. (2001). They have (2001) described a 3–dimensional structure of PS I through determination of x-ray structure from thermophilic cyanobacterium (*Synechococcus elongatus*) at 2.5 Å resolution. According to these authors, the x-ray crystal structure of PS I from this bacterium gave a detailed atomic picture of the 12 protien subunits. Further, the cofactors — including 96 chlorophylls, 2 phylloquinones, 3 Fe 4S4 clusters, 22 carotenoids, 4 lipids, a calcium ion (Ca^{2+}) and 201 water molecules — were also revealed. As is already known, PS I and PS II belong to the two classes of reaction centres based on the terminal electron acceptors. Type I ($Fe_4 S_4$ clusters) and Type II (quinones) comprise the two classes. The purple bacterial RC and PS II belong to the type II RC. Three-dimensional structures are known at high resolution (Lancaster et al., 2000; Zouni et al., 2001). Earlier, a structural model for cyanobacterial PS II at 4A° resolution was available. However, further detailed structural information has been lacking till recently.

PS I from cyanobacteria, when isolated, exists in a trimer form (relative molecular mass 3 x 356000). Each monomer has minimum of 11 different protein subunits which coordinate over a hundred co-factors. There is a large internal antenna system for light harvesting by PS I. The primary charge separation takes place followed by passing through the electron.

P 700 → Ao (Chla) → A_1 (Phylloquinone) → $Fe_4 S_4$ Clusters, Fx→ F_A and F_B (Brettel, 1997). At the stromal side, the electron from F_B is passed on to ferredoxin and then or to NADP. P 700^+ is reduced by plastocyanin at the lumenal side (Jordan et al., (2001). PS I, in its trimeric form, is clove leaf shaped. Its rotation axis coincides with the C3 axis. At the trimerization domain close to C3 axis, PaL forms most of contact between the monomers (Jordan et al., 2001).

Nine protein subunits with transmembrane α helices (PsaA, PsaB, PsaF, PsaI, PsaJ, PsaK, PsaL, PsaN and PsaX) and three stromal subunits (PsaC, PsaB, PsaE) are present in PS I besides the subunits PsaA and PsaB located at the centre of the PS I monomer. Most of the antennae Chl a molecules and carotenoids are bound to Psa A/B. A docking side for plastocyanin (electron donor to PS I) at the lumenal side and the ferredoxin (electron acceptor from PS I) at the stromal side are present. The subunits C, B and E bind the ferredoxin component.

The subunits of PsaA, PsaB contain eleven transmembrane helices each. These helices are divided into an amino terminal domain (6 α helices; A/B–a to A/B–f) and carboxyl terminal domain (5 α helices; A/B–g to A/B–k). The lumenal surface PS I is formed by loop regions connecting transmembrane α helices of subunits PsaA and PsaB. Subunit PsaF contributes prominent structural features to this surface of PS I. The loops at the stromal surface of PsaA/B constitute the binding intersurface to subunits PsaC, PsaD and PsaE. They are close together in an arrangement shaped like a crescent. PsaC contains two Fe_4S_4 clusters, F_A and F_B. PsaE consists of 5-stranded β barrel. The small subunits, along with the PsaA and PsaB, bind the cofactors of core antennae system. PsaJ, PsaK, PsaL, PsaM and PsaX coordinate the magnesium ions of antenna Chla molecules. PsaF, PsaI, PsaJ, PsaL, PsaM and PsaK are in hydrophobic contact with carotenoids (Jordan et al., 2001).

The principal component of PS I in its function is formed by six chlorophylls, two phylloquinones and three Fe_4S_4 clusters. Chlorophylls and phylloquinones are on two, branches A and B. The phylloquinones might correspond to electron acceptor A_1. Whether one or both branches of ETC are active in PS I is still controversial. The assignment of the two Fe_4S_4 clusters in PsaC to F_A and F_B is now clear. The arrangement of the clusters with F_A being closer to Fx than Fb suggests a sequence, Fx to F_A to F_B.

The core antenna at PS I is formed by 90 Chl a molecules and 22 carotenoids. Of the antenna Chl a molecules, 79 are bound to PsaA and PsaB. Small subunits, namely PsaJ, PsaK, PsaL, PsaM, PsaX, and lipid also coordinate Mg^{2+} of 11 Chl a molecules. The antenna is extended by two sets of 18 peripheral Chl a molecules bound by PsaA and PsaB.

Twenty-two carotenoids are identified and modelled as β carotene, 16 of them are all – trans – configuration, 5 contain 1 or 2 cis bonds. They might serve two major functions: (a) light harvesting in the 415 – 570 nm range where Chl a absorbtion is negligible, and (b) Photoprotection by quenching excited Chl a triplet states.

Four lipids are associated with PsaA/PsaB, three are phospholipids (I, III, IV) and one is a galactolipid (II). The head groups of lipids are located on the stromal side and form hydrogen bonds with PsaA./PsaB. The lipid III binds the antenna chlorophyll. The functional importance of lipids is brought out by the fact that they are located close to the core of PS I (Jordan et al., 2001).

The crucial part of PS I comprises the three subunits PsaA, PsaB and PsaC (PS I-A, PS I-B and PS I-C) that together bind the electron acceptors. The subunits are encoded by the chloroplast genome (Scheller et al., 2001).

The PS I RC polypeptides, PsaA and PsaB are encoded by psaA and psaB genes. These two proteins are hydrophobic and contain as many as 11 transmembrane segments in the heart of the PS I reaction centre and have a role in binding PsaA and PsaB proteins together. The region may also contain other cofactors and 3 histidine residues in helix XIII which may bind P 700 or Ao. PS I-C binds the terminal acceptors F_A and F_B.

The clearest evidence for the identity of P 700 as a dimer is provided by specialized spectroscopic studies. In the ground state, P 700 is a dimer and the P 700+ cation is localized on one of the two molecules that comprise the chlorophyll special pair. Potential of Ao/Ao- pair has not been measured directly but indirect evidence indicates a value of − 1.01 V. Ao is a monomer of chlorophyll anion. The redox potential of A_1, secondary acceptor is − 810 mv. The double reduction of phylloquinone traps the electron in a high potential energy sink, thereby preventing the flow of electron to Fx or back to Ao. This behaviour explains a proposed photoinhibition of PS I in high light. The reconstitution of phylloquinone has been demonstrated (Golbeck, 1992).

Redox potential of Fx cluster is − 730 mv, making it the lowest potential known biological iron–sulphur cluster. The identification of effects as a 4 Fe–4S cluster dictates the minimum polypeptide architecture of the PS I complex. Since ligands to the Fx cluster are cystines and as there are only three conserved cystine residues on the PsaA and two conserved cystine residues on PsaB, Fx is perhaps an inter polypeptide cluster ligated between homologous cystine residues on the PsaA and PsaB heterodimer. The role of Fx iron–sulphur cluster in the forward electron flow is not clear. However, most researchers assume that the sequence of electron flow is $A_0 \rightarrow A_1 \rightarrow F_1 \rightarrow F_A/F_B$.

1.6B. Core Complex of Photosystem I

PS I core contains several integral low molecular mass proteins, PsaI, PsaJ, PsaK, PsaL. PsaI and psaJ genes are located in chloroplast genome and serve an important function in the primary processes. These polypeptides have masses 3.9 and 4.1 in spinach and 4.1 kDa for both in pea. They are both hydrophobic with one α - helix; PsaK is nuclear encoded. PsaG and PsaK are integral membrane proteins of 11 and 9 kDa, respectively. It has been shown PS I-G and PS I-K are involved in interaction with LHC I in higher plants. In the absence of PS I-K, the content of Lhca 2 and Lhca 3 decrease (Scheller et al. 2001). The subunit composition and organization of the PS I complex from higher plants (Scheller et al., 2001) is shown in Figure 1.7. PS I core may also contain PsaG and PsaM, PsaN in eukaryotic systems. PsaG is not found in cyanobacteria; PsaL is nuclear encoded and has mass of 18 kDa in higher plants. PsaM has been found on the chloroplast genome of algae. Its role is presently unknown (Scheller et al., 2001).

PS I complex contains several extrinsic proteins and, like PsaC, PsaD, PsaE and possibly PsaH in eukaryotes, is located on the stromal side of the membrane. PsaC gene is located in the chloroplast genome. It has a mass of 8.5 kDa and is highly acidic. In *Synechocystis*, it has two regions characterstic of 8Fe − 8S proteins. In higher plants at position 37, a

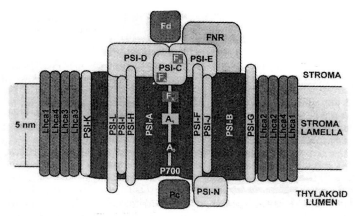

Fig. 1.7. Schematic diagram for the organization of eukaryotic photosystem I supercomplex. The relative positions of the subunits of the core (PS I – A to PS I – N) and of light harvesting complex (Lhca1 to Lhca 4) are shown. The location of P700 and the electron acceptors on the PS I A/B heterodimer is depicted. (Figure reprinted with permission by Elsevier Ltd from HV Scheller et al (2001) Role of subunits in eukaryotic photosystem I. Biochem. Biophys Acta 1507 pp.41-60. Copyright 2001 Elsevier Science B.V.)

tryptophan – histidine pair, in dicots and glycine – proline in monocots are present. The PsaC protein functions in locating FA and FB terminal electron acceptors in PS I. The redox properties of F_A and F_B clusters (4 Fe-Fs) are conferred predominantly by primary sequence of PsaC protein, but not by interaction with PS I core. The PsaC gene from *Chlamydomonas*, when inactivated, results in a destabilization of PS I complex, so that none of the proteins accumulate. In general, PsaC protein is found to be essential for the assembly of PS I complex.

PsaD protein is nuclear encoded and known to interact strongly with ferredoxin. It is positively charged with a mass of 17.9 kDa in eukaryotes. This protein has two known roles in PS I : (see also Lagoutte et al (2001))

1. It acts as a docking protien for soluble 2 Fe – 2S ferredoxin to the PS I complex. The need for a docking is derived from the requirement that iron – sulphur cluster in ferredoxin is properly aligned with F_A or F_B iron – sulphur cluster in PsaC for efficient electron flow.
2. The second role for PsaD is to promote orientation of PsaC, leading to a stable PS II complex. PsaC binds loosely to PS I core in the absence of PsaD. Accordingly, the function of PsaD resides in ferredoxin in docking or PsaC binding (or both). A third function of PsaD protein is to stabilize the PsaE protein on PS I core. PsaE gene is nuclear coded in eukaryotes and is 9.7 kDa in spinach. PsaE protein binds to PsI core simultaneously

with C and D. Lack of PsaE leads to high susceptibility to photodamage of PS I complex. PsaH is nuclear encoded and has a mass of 10.2 kDa in spinach and barley. Although the function is not known, it might play a role in docking the LHCs to PS I core.

PsaC, PsaD, PsaE are in intimate contact with each other on the stromal side of PS I complex.

PsaF and PsaN are the two known proteins (Scheller, 2001) located in the lumenal side of the membrane. The PsaF gene is located on the nuclear genome and is of 18 kDa. The location of PsaF protein on the lumenal side is consistent with its role in modulating plastocyanin and Cyt c 553. In function, this protein has the role of plastocyanin docking or as light-harvesting chlorophyll protein. PS I-N is a small subunit of about 10 kDa.

Antennae proteins of PS I holocomplex contain, in eukaryotes, a set of light harvesting chl proteins that increase the optical cross section due to 100 additional chlorphyll molecules. The polypeptide composition of LHC I complexes has been determined by the molecular genetics approach. LHC I proteins are now known to be members of a larger family which includes LHC II and other minor complexes in PS II. The peripheral antennae of higher plant PS I (LHC I) is composed of four proteins encoded by nuclear genes, Lhca1, Lhca2, Lhca3 and Lhca4 with masses 20-24 kDa (Golbeck, 1992).

A comparison of the antenna system of PS I with that of PS II may be useful at this point. The peripheral antennae Chla molecule bound to PsaA and PsaB in PS I resemble that of integral antennae proteins, CP 43 and CP 47 of PS II. The PS II reaction centre of PS II formed by D1 and D2 is structurally similar to C-terminal domains of PsaA and PsaB. However, unlike PS II, the C-terminal domains of PsaA and PsaB cannot be considered as pure RC domains of PS I. Also, the central part of PS I core antennae containing 43 Chl is different from the corresponding region in PS II. Each Chl a molecule in the central PsaA/PsaB antennae of PS I is at a relatively short distance from ETC, and the transfer of excitation energy is highly efficient. A complete lack of such a central antenna in PS II may be of lower efficiency compare to PS I in transferring excitation energy to RC Jordan et al. (2001).

The structure of Photosystem II and Photosystem I described here is by no means an exhaustive account. It is primarily intended to provide a background information in order to facilitate the understanding of the phenomena of photoinhibition, photoprotection and photosynthetic acclimation that might involve certain structural alterations in addition to the physiological and biochemical parameters. Appropriate references are provided from which the reader can obtain full details.

2
Photoinhibition

2.1. INTRODUCTION

In general, photoinhibition is regarded as the light-dependent loss of photosynthetic efficiency, normally occurring under conditions of light harvesting antennae absorbing more excitation energy than can be dissipated by photochemistry of photosynthesis (Powles, 1984; Krause, 1988; Osmond, 1994; Long et al., 1994; Hurry et al., 1997). The primary target of damage during photoinhition is photosystem II (PS II). Further, it is also shown that specifically, the D1 subunit of PS II reaction centre is degraded, a significant consequence of photoinhibition (Ohad et al., 1984; Adir et al., 1990). The D1 protein of PS II has the highest turnover rate, particularly at high light intensities and may even be 50 to 80 fold higher than any other thylakoid proteins (Prasil et al., 1992; Andersson et al., 1994). Though the primary target of photoinhibition is photosystem II, the photoinhibition of photosystem I (PS I) is not altogether unknown. This is evident under conditions of illumination of cold sensitive plants at low temperatures. Terashima et al., (1994) and Terashima et al., (1998) have reported the inhibition of PS I in *Cucumis* and subsequently also in other plants such as potato (Havaux and Davaud, 1994), cucumber (Sonoike 1995), barley (Tjus et al., 1999) and spinach (Tjus et al., 2001). However, PS I photoinhibition has been much less studied (Hihara and Sonoike, 2001). Looking at a different perspective, Kato et al., (2002) have referred photoinhibition to a net inactivation of PS II which can be ascertained by a balance between gross inactivation (photoinactivation) and the simultaneous recovery of PS II through the process of D1 protein turnover.

Over the years, the usage of the term photoinhibition has undergone several modifications from its classical definition by Kok, (1956), who considered the phenomenon as a decrease in the rate of photosynthesis under excessive illumination. Gradually, it became more or less a synonym for photoinhibition of PS II, particularly when associated with a reduction in overall photosynthetic capacity (Long et al., 1994; Alves et al., 2002).

Considerable ambiguity exists in the understanding of photoinhibition as used by different researchers (Krause, 1988; Critchley, 1998). The existence of uncertainity in terminology has also been raised by Werner et al. (2002). A concept of dynamic and chronic photoinhibition has been developed by Osmond (1994). Dynamic photoinhibition is ascribed to Δ pH dependent fluoresence quenching (qE), a major component of qN i.e. NPQ. This is a rapidly-relaxing phenomenon, while chronic photoinhibition is a slow-relaxing system and involves reduced light saturated rates of photosynthesis. Chronic photoinhibition leads to irreversible photodamage to PS II and, therefore, resumption of normal photochemistry requires *de novo* synthesis of the proteins of reaction centre (Owens, 1994; Young and Frank, 1996; Demmig-Adams and Adams, 1996b). Photoinhibition is regarded to be a two-component process; one of them involving a short-term (less than one hour) photoprotection and the other a long-term photodamage (several hours) consisting of degradation and replacement of photodamaged D1 protein (Kitao et al., 2000). These two components obviously correspond to the dynamic and chronic photoinhibition. In this context, Werner et al., (2002) have used this concept of dynamic and chronic photoinhibition for evaluation of an ecosystem performance under natural conditions. They have based their definition on the recovery time of Fv/Fm, a measure of photochemical efficiency of PS II.

Andersson and Aro (2001), while reviewing the vulnerability of PS II and the turnover of D1 protein, considered avoiding the dispute of the usage of photoinhibition and confined themselves to an irreversible damage in the D1 reaction centre protein.

Plants must maintain an effective balance between energy supply and energy consumption. Plants possess several photoprotective measures to ensure such a balance, particularly through safe dissipation of excess energy (Demmig – Adams and Adams, 2000). Such non-photochemical dissipation of energy affords photoprotection through a thermal sink for absorbed photons (Osmond et al., 1997). A photosynthetic apparatus rapidly adjusts to changing external environment to avoid any dishormony between energy influx and the capacity for its utilization, thus maintaining homeostasis of the photosynthetic processes (Nedbal and Brezina, 2002). However, photoinhibition does occur under conditions of inadequate photoprotection. Photoinhibition in vivo occurs as a result of excessive excitation pressure on PS II with Q_A being mainly in the reduced level due to sustained high visible light. It is considered that tolerance to photoinhibition represents a response to PS II excitation pressure and also that algae and higher plants adjust in different ways in this respect (Gray et al., 1996). Over-reduction of Q_A is not a prerequisite for in vivo photoinhibition. It is important to consider the opposing effects of photoinactivation and photoprotection. Regulation of PS II and D1 protein turnover are extremely important areas for plant survival, even under low light. Under limiting to saturating light, D1 protein turnover is fast enough to prevent net photoinactivation of PS II in vivo. D1 protein turnover is linearly related to maximal photosynthesis upto light saturation, but above saturating light, non-functional PS II centres with intact D1 protein accumulate and D1 protein turnover is no longer linearly related to irradiance. Hence, D1 protein turnover is a major photoprotective strategy for PS II (Anderson et al., 1997a).

It is believed that the avoidance of photoinhibition at peak light conditions requires an investment in a photosynthetic apparatus of a high capacity. Such a system is expensive to create as also maintain (Ogren, 1994). Accordingly, a strategy with many leaves of rather moderately high photosynthetic capacities matching average light conditions should perhaps be more beneficial than a strategy having fewer leaves of peak light capacities (Ogren, 1994). In certain situations, leaf orientation such as the flag leaf of field-grown rice adjusting to reduced photoinhibition has been reported (Murchie et al., 1999).

It can also be argued that evolution has produced a system of photosynthesis with high efficiency at limiting light while maintaining mechanisms for mitigation of photodamage at high light intensities (Ort, 2001).

2.2. OCCURRENCE OF PHOTOINHIBITION

The phenomenon of photoinhibition is supposed to occur in all oxygenic photosynthetic organisms in vivo. The process can also be found in isolated thylakoids in vitro. (Tyystjarvi et al., 1999a). Typically, photoinhibition has the potential for the inhibition of photosynthesis and reduced plant growth (Melis 1999). PS II damage and repair cycle is highly relevant for the photosynthetic function in view of the fact that when the rate of photodamage exceeds the rate of repair, the photoinhibition of photosynthesis is apparent. The events in PS II damage and repair cycle occur in a sequential manner. They can be depicted as partial disassembly of PS II complex, photodamage to PS II core, followed by a migration to chloroplast stroma from the granal region. The process continues through D1 degradation, D1 biosynthesis, the insertion of new D1 protein into the thylakoid membrane and, finally, by reassembly of PS II complex containing functional D1/D2 heterodimer with electron transport (Melis, 1999).

In view of the fact that photosystem II is the most susceptible component of the chloroplast for light stress, it is relevant to briefly recapitulate (details in Chapter I) the overall structural complexity of the system in understanding the process of photoinhibition. The reaction centre of PS II contains six chlorphyll molecules, two pheophytins, one non-heme iron and two quinones Q_A and Q_B bound to the heterodimer, D1 and D2 polypeptides (Ke, 2000b). Therefore, it is the D1/D2 heterodimer that binds all essential redox components of PS II. The binding of Q_B is to the D1 protein, while Q_A is to the D2 protein. The structure of PS II polypeptides and their transmembrane helices has been studied through electron crystallography at 8°A by Rhee et al., (1998) and by Zouni et al., (2001). The D1 and D2 subunits along with cytochrome b-559 and psb I and psb W gene products are referred to as PS II RC (Nanba and Satoh 1987; Fotinou and Ghanotakis, 1990; Psylinakis et al., 2002). The core antennae consisting of 43 and 47 kDa polypeptides are present on the opposite sides of the RC (Review by Tracewell et al., 2001). Further, the PS II complexes retaining the oxygen-evolving ability possess — in addition to the CP 43 and CP 47 — an oxygen-evolving complex containing 33, 23 and 17 kDa proteins (Hankamer et al., 2001) and several low molecular mass subunits. This constitutes PS II core complex (Ghanotakis and Yocum, 1986; Psylinakis et al., 2002).

The very rapid and light-dependent turnover of D1 polypeptide had been known for some time. A half time for turnover can be as short as 60 minutes. PsbA gene codes for this protein and mRNA produced by this gene are in abundant transcripts. D1 is also remarkable since it ligates all important sites of PS II electron transfer. The D1 not only binds Q_B but also the primary electron donor and acceptor, P 680 and pheophytin. Additionally, it also contains tyrosine residue at position 161, usually denoted as Z which is a redox intermediate between Mn cluster and the P 680+/Pheo-radical pair. D1 also acts as a ligand for binding Mn cluster, making it unique and there is apparently no other protein in PS II that carries such functional responsibility (Barber and Andersson, 1992).

2.3. PHOTOINHIBITION OF PHOTOSYSTEM II

Light energy is essential for photosynthesis but excessive light is detrimental to the photosynthetic activity. Such a phenomenon is known as photoinhibition, which results in the destruction of pigment molecules and degradative cleavage of D1 protein. The damage due to excess light is corrected by replacement of the degraded components through renewed synthesis and reassembling of PS II RC. Photoinhibition is observed when the repair is less than the damage (Ke, 2000). It has also been shown that PS II is more prone to photoinhibition in state 2 light conditions (Finazzi et al., 2001). High light stress and recovery were studied in *Dunaliella salina* (Masuda et al., 2002) and it was observed that chlorophyll antennae size of PS II is post-transcriptionally regulated by the availability of chlorophyll.

Under high light intensity, it is possible for Q_A to undergo double reduction and, therefore, be unable to accept electron from newly-formed pheophytin. Under such conditions, the separated charges P 680$^+$, pheo$^-$ recombine. This recombination results in the formation of triplet ^{3}P680. The life time of the ^{3}P680 is long enough to react with triplet O$_2$ in order to form singlet oxygen (Ke, 2000b).

Photoinhibition may follow either of the two pathways — the acceptor side and the donor side. The mechanism of photoinhibition can be elucidated in vitro with the D1/D2/cytb 559 complex. Several in vitro studies on the photoinactivation have resulted in the identification of two different pathways for the photodamage of PS II during photoinhibition (Barber and Andersson, 1992; Prasil et al., 1992). These pathways are understood to occur either on the acceptor (acceptor side mechanism) or the donor side (donor side mechanism) of PS II. They were initially elucidated by Telfer and Barber (1994), making use of the PS II RC complex, D1/D2/cytochrome b-559.

2.3A. Acceptor Side Photoinhibition

The acceptor side photoinhibition is caused by an accumulation of reduced Q_A- species leading to the formation of chlorophyll triplets and singlet oxygen, which in turn, damage the D1 protein (Figure 2.1). On the other hand, the donor side photoinhibition is caused by relatively long lived P 680+ (Ke, 2000b). This donor side impairment is due to the rate of electron donation to PS II RC not matching with the electron removal. The life span of P 680$^+$ and TynZ$^+$ are said to increase under such conditions.

The site of inactivation of electron transport is Q_A, while Q_B site was found to be necessary for the degradation of damaged D1 (Andersson and Barber, 1996). Vass et al., (1992) have used EPR and fluorescence measurements with isolated thylakoids in order to understand the mechanism of acceptor side photoinhibition. Q_A is found to reduce to the semiquinone Q_A-. This is reoxidized by Q_B. Due to blockage of electron transfer from Q_A- to Q_B, which happens due to overreduction of plastoquinone pool, Q_A- is double reduced and protonated, leading to the formation of Q_AH_2 (Styring and Jegerschold, 1994). Mulo et al., (1998) using mutants with deletions in D-E loop of D1 from *Synechocystis*, have suggested the existence of interacting factors in the regulation of D1 turnover but not exclusively dependent on light intensity.

This Q_AH_2 is then dissociated from its binding site, leading to the charge recombination between P680$^+$ (oxidized form of chlorophyll primary electron donor) and the reduced pheo$^-$ (reduced form of primary electron acceptor, pheophytin). This process of recombination of primary radical pair, P 680+ pheo$^-$ results in the production of chlorophyll triplets. Chlorophyll triplets in general could either be produced by intersystem crossing from a singlet excited state or by the recombination of the charge

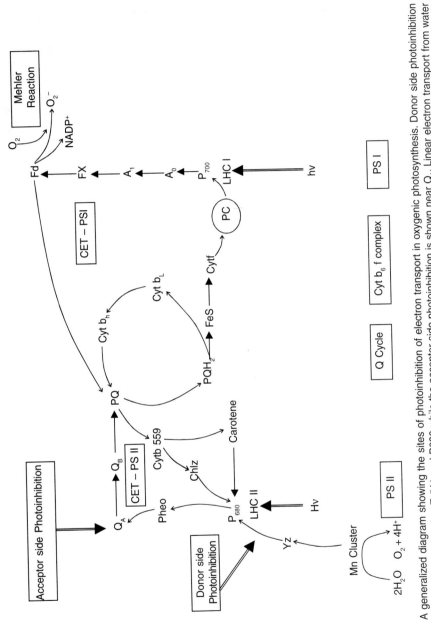

Fig. 2.1. A generalized diagram showing the sites of photoinhibition of electron transport in oxygenic photosynthesis. Donor side photoinhibition occurs between tyrosine Z (Y_Z) and P680 while the acceptor side photoinhibition is shown near Q_A. Linear electron transport from water to $NADP^+$ and the cyclic electron transport around PS I (CET-PS I) and PS II (CET – PS II) are indicated.

separated radical pair P680$^+$ Pheo$^-$ (Santabarbara et al. (2002a). The rate of intersystem crossing is significant in the isolated pigment protein complexes. The PS II reaction centre recombination triplet (Rutherford et al., 1981) is significantly produced when QA is doubly reduced and in the form of uncharged QAH$_2$. Santabarbara et al., (2002b) have demonstrated that chlorophyll triplets could also arise from chlorophylls not associated with antennae complexes but from those that are weakly coupled or uncoupled complexes. These complexes do not transfer energy to RC of PS II. Triplets are formed on these complexes which result in singlet oxygen, leading to photoinhibition. It was also shown that a much higher yield of triplets arise than the recombination triplets. According to them, the recombination triplet formation is not an exclusive pathway for chlorophyll triplet production. Santabarbara et al., (2002b) have further shown — confirming their earlier findings (Santabarbara et al., 1999, Santabarbara et al., 2001a,b) and of Tyystjarvi et al., (1999) — that a substantial component of photoinhibition is not related to the light absorption by PS II antennae but is obtained due to absorption by a small number of chlorophylls uncoupled from the antennae matrix. This view — at variance from the current thinking on the formation of chlorophyll triplets, and the consequent singlet oxygen production leading to photoinhibition — requires further analysis and confirmation. While triplet chlorophyll is not by itself toxic but readily reacts with triplet oxygen in producing a highly damaging oxygen species, a singlet oxygen is as follows:

$$^3Chl + {}^3O_2 \dashrightarrow {}^1Chl + {}^1O_2$$

The singlet oxygen is said to destroy chlorophyll P 680 and also interacts with the amino acid residues of protein matrix (Telfer et al., 1994) in both D1 and D2, as shown in studies by Sharma et al., (1997). It is also known that the reaction centre can be photodamaged by other toxic oxygen and hydroxyl radicals. Tjus et al., (2001) have shown that superoxide and H_2O_2 can also induce degradation of D1 protein. The site of damage in the acceptor side photoinhibition is the stroma-exposed DE-loop of D1 protein. It has now become evident that D1 protein of RC of PS II participates in the ligation of the redox active RC components (Trebst, 1986) and, therefore, its rapid turnover has become an important area of photosynthesis research. Q_A is apparently the site of inactivation of the electron transport, while Q_B site is found to be essential for triggering the damaged D1 protein.

2.3B. Donor Side Mechanism

The donor side photoinhibition by damage to D1 protein is caused by long-lived oxidative species P 680+ and the oxidized tyrosine electron donor Tyr z$^+$ (Jegerschold et al., 1990; Yamamoto, 2001; Andersson and Aro, 2001). P 680$^+$ and Tyr z are long lived either when manganese cluster is lost from PS II or under conditions of low chloride content. The photodamage to D1 occurs in an oxygen- independent route, in the donor side of PS II (Andersson and Barber, 1996). Probably, the donor side inactivation takes place at a site between manganese cluster and P 680 (Eckert et al. 1991) (Figure 2.1).

Blubaugh et al., (1991) studied the donor side photoinhibition using EPR and optical spectrophotometric analysis and found that the order of susceptibility of PS II components was chlorophylls/ carotenoids > Tyr z > Tyr D >> P 680, Pheophytin, Q_A. It is not fully established whether oxygen-independent degradation of D1 protein is characterstic of the donor side mechanism or if oxygen radicals are also involved in this. In donor side mechanism, the damage occurs at the lumen-exposed AB loop, while it was mentioned that in the acceptor side inhibition, the damage happens at the stroma-exposed DE loop.

Most of the studies on photoinhibition were made using in vitro systems. However, it is still uncertain regarding the cause of photoinhibition in vivo. Kettunen et al., (1996) have studied light- stressed D1 protein degradation in pumpkin leaves and found it to be similar to that of donor side mechanism demonstrated in vitro. This suggested that photoinhibition takes place through donor side inhibition in vivo.

2.4. PHOTOINACTIVATION

Photoinactivation is considered to be a light-induced loss of charge separation and is accompanied by damage and replacement of PS II D1 protein, leading to restoration of normal photosynthesis (Ohad et al., 2000). PS II photoinactivation results from excessive light energy which is neither consumed for photosynthetic electron transport nor dissipated as heat (Kato et al., 2003).

The PQ pool becomes fully reduced leaving Q_B site. This makes PQ tightly bound to Q_A site and accumulates in a long-lived singly reduced state Q_A-. This long-lived semiquinone becomes protonated to Q_A–H+. Such semiquinone receives a second electron, resulting in the formation of abnormal $Q_A H_2$. The study described above (Vass et al., 1992) also provided the connection between the photoinactivation to photodamage to RC and subsequent degradation of D1 protein. It was also considered that photo-oxidation of Cyt b559 *via* oxidation of Chl z leads to the photoinactivation of the PS II reaction centre (Koulougliotis et al., 1994). Photoinhibition leads to accumulation of photochemically-inactive PS II reaction centres (Krause, 1994; Melis, 1999). These inactive centres are also energy-dissipating structures. The reactivation involves turnover of

D1 protein of the reaction centre. Thus, the formation of energy dissipation PS II centres and their recovery suggests that photoinhibition is rather a protective mechanism for down regulation of photosynthesis to avoid stress caused by high light (Krause, 1994).

Functional PS II complexes were lost with increase in cumulative light dose (Photon exposure) in *Capsicum annuum* leaf discs (Lee et al. 1999). It was also made clear that photoinactivation of PS II is related to light dose (irradiance and duration of illumination). Further, Lee et al., (2001) have demonstrated, using *Capsicum annuum* leaves, that photoinactivated PS II complexes actually afford photoprotection of functional neighbours. Similarly, photoprotection of functional PS II by inactive centres has been demonstrated in high light acclimated grapevines by Flexas et al. (2001). The findings of Lee et al., (2001) support the work of Oquist et al., (1992) and of Russel et al., (1995) on the role of photoinactivated complexes. Extending their work, Lee et al., (2002) using *Capsicum* leaves have suggested that the photoprotection of functional PS II neighbours requires thylakoid lumen acidification and also that an optimal pH of stroma is critical for D1 protein synthesis for recovery from PS II photoinactivation.

As stated above, the chlorophyll triplet arises from recombination of P 680 + /Pheo$^-$, which is produced by impairment of the Q_A function. ^{3}Chl will then react with molecular ($^3O_2$) oxygen to form singlet oxygen. Singlet oxygen ($^1O_2$) is highly reactive and can oxidize nearby pigments and redox components. It was concluded, therefore, that singlet oxygen is the primary damaging species to PS II RC during the acceptor side photoinhibition (Vass et al., 1992). In contrast to this view, Anderson, et al., (1997; 1998) believe that it is P 680$^+$ which is the damaging speices, but not the singlet oxygen ($^1O_2$). Other toxic oxygen and hydroxyl radicals may also be acting as damaging species. Photoinactivation is an inevitable consequence of photosynthesis. PS II inactivation occurs at the rate of one PS II with absorption of 10^6 to 10^7 photons (Anderson et al., 1997a). Since photoinactivation is a concurrent feature with photosynthesis, photoinhibition is determined by the balance between photoinactivation and concurrent recovery through *de novo* synthesis of D1 (Greer et al., 1986; Aro et al., 1993; Kato et al., 2002). Photodamage to PS II occurs with a half time ranging from 8h to 30 min., while the repair process with a half time of 60 min. (Melis, 1999) Photoinactivation in vivo occurs at all light levels from limiting to saturating to supersaturating (Baker, 1996). Leaf photosynthetic capacity is maintained only when the rate of D1 protein degradation is matched by the rate of repair process (Baker, 1996). However, the rate of D1 protein turnover is swift and prevents net loss of PS II function. Under high light, non-functional PS II complexes (still possessing D1 protein) accumulate (Anderson and Aro 1994). Jahns et al.,

(2000) have proved the protective role for zeaxanthin against high light inactivation of photosystem II in *Chlamydomonas* leading to a slower D1 protein turnover.

It was also established that photoinactivation of PS II depends on the antennae size in light acclimated pea leaves by Park et al., (1997) who used the target theory (Sinclair et al., 1996) for the study of photoinactivation in vivo.

2.5. DEGRADATION OF D1 PROTEIN

Once the damage to D1 protein has occurred, the next step is the degradation of the protein through proteolysis. Photosystem II exists as a dimer and it is active only in this form (Hankamer et al., 1997b). Prior to D1 degradation, it is observed that PS II dimers turn into monomers which migrate to the stroma from the granal region (Gonzalez et al., 1999; Cai et al., 2002). It was observed recently (Cai et al., 2002) that there was no monomerization of PS II in soybean leaves under photoinhibition induced by saturating irradiance. This lack of monomerization is interpreted as related to the chromic photoinhibition rather than to the dynamic photoinhibition (Cai et al., 2002).

The pattern of fragmentation of D1 protein is dependent on the mechanism of photoinhibition. Following the acceptor side photoinhibition, the cleavage of D1 protein takes place at the stromal DE loop and the cleavage results in a 24 kDa N-terminal fragment and 10 kDa C-terminal smaller fragment. When the donor side photoinhibition occurs in PS II particles, the fragmentation of D1 occurs at the lumenal AB loop resulting in 24 kDa C-terminal and 10 kDa N-terminal fragment (Andersson and Barber, 1996; De Las Rivas et al., 1992).

It can be remembered that D1 protein is a intergral membrane protein consisting of five transmembrane helices (A to E) with N-terminus exposed to the stromal and C-terminus exposed to lumenal surface of thylakoids (Hankamer and Barber 1997). It is obvious that the heterodimer formation, along with D2 protein surrounded by as many as 25 subunits, results in a super complex of PS II. Accordingly, the degradation within the complex is a highly specialized process involving specific proteases.

It is probably quite important that certain conformational changes in PS II are required to trigger D1 protein degradation. Several events are involved which occur in sequence during photoinactivation of electron transport and D1 protein turnover. This sequence of events is referred to as PS II damage and repair cycle (Adir et al., 1990; Barber and Andersson, 1992; Marshall, 2000, Yokthongwattana et al., 2001). Photodamage occurs due to an imbalance in the electron transfer, leading to production of toxic intermediates. The sequence of events likely to occur are understood as follows: Proteolysis and removal, insertion of new protein copy and restoration of functional PS II RC. These events can be initiated either from the acceptor or donor side.

2.5A. Proteolysis of Damaged D1 Protein

This process involves a proteolytic enzyme system which can distinguish between the functional and photodamaged RC (reviewed in Andersson and Aro, 2001). It is achieved by an irreversible photodamage inducing a triggering of RC D1 protein which becomes the substrate for proteolysis.

Hankamer et al., (1997) reviewed the structure and organization of photosystem II. D1 has five transmembrane helices (A to E) with a N-terminus and C-terminus exposed to stromal and lumenal surfaces respectively. The structure of D1 is unique and highly complex, surrounded by 25 other subunits within the PS II super complex.

Plants survive photoinhibition through an intricate repair mechanism consisting of proteolytic degradation and replacement of damaged PS II RC D1 protein. The proteolytic process is less known in the key step of the overall repair process which is the degradation of damaged D1. A 23 kDa N-terminal fragment and later a 10 kDa C-terminal have been identified. This pattern indicates that cleavage occurs in the stromal loop between D and E helices exposed at outer thylakoid surface. Primary cleavage of D1 protein is a GTP dependent process, whereas secondary cleavage steps require ATP and Zn.

A general model for proteolytic events is as follows (Spetea et al. 1999).

$$D_1 \text{ protein} \xrightarrow[\text{Serine-type protease}]{\text{Mg GTP}} 23 \text{ kDa N-terminal} \xrightarrow[\text{FtsH}]{\text{Mg ATP, Zn}^{2+}} \text{low MW fragments}$$

The chloroplast proteases are known to be homologous to bacterial enzymes including ClpA and FtsH (Lindhal et al. 1996). Spetia et al., (1999) have concluded from their work that FtsH is responsible for the secondary proteolysis of D1 degradation.

The identification of D1 protein fragments is an intricate process. The total degradation of D1 protein might produce a large number of smaller fragments. However, an understanding of the initial cleavage of protein requires precise identification of the primary fragments around the Q_B site of D1 protein, which is the primary cleavage site identified as early as 1987 by Greenberg and associates (Aro et al., 1993a,b; Andersson and Barber 1996). The primary cleavage is located in the stromal DE loop. The fragments are N-terminal 23 kDa and C-terminal 10 kDa. Additional 16 kDa fragment is perhaps due to a cleavage in the lumenal loop, between C and D helices. This pattern is applicable principally to the aceeptor side photoinhibition. On the contrary, the D1 degradation under donor side induced photoinhibition leads to an N-terminal 9 kDa and C-terminal 24 kDa fragments (De Las Rivas et al. 1993; Andersson and Barber, 1996; Andersson and Aro, 2001). Accordingly, the site of primary cleavage under the donor side is not the stromal DE loop, as is in the case for acceptor side photoinhibition and degradation phenomenon, but it is the lumenal loop between the A and B helices.

The proteases involved in the degradation of D1 protein of PS II subsequent to its damage in photoinhibition are crucial to the process. Different proteases have been identified in photosynthetic organisms that are associated with D1 turnover or assembly (Adam et al., 2001; Bailey et al., 2002; Zheng, et al., 2002). These proteases are homologous to the proteases of enfacteria. Generally, three kinds of proteases are recognized in chloroplasts (Adam et al., 2001). One of these is the stromal protease DegP, which is extrinsically attached to the lumenal and stromal surfaces of thylakoid and has been shown to be involved in D1 cleavage in vitro assays (Haussuhl et al., 2001; Bailey et al., 2002). A second type of protease, FtsH, is also attached to the thylakoid membrane. Another type is the soluble Clp protease which has also separate chaperone and proteolytic subunits. The Clp protease is the newly-dentified proteolytic system which targets specific polypeptide substrates (Zheng et al., 2002). Zheng et al., (2002) have recently shown that high light intensity and cold treatment produces substantial increases in chloroplast Clp proteins such as ClpD and ClpP. This finding suggested that Clp proteins play a role in protein turnover associated with high light and cold temperature. However, it still needs to be appreciated that the precise role of the different proteolytic systems is not fully understood (Zheng et al., 2002). The proteases which are associated with thylakoid membrane were found to be of serine type (De las Rivas et al., 1993). Recently, Lindhal et al., (2000) and Haussuhl et al., (2001) using biochemical and genomic analysis of bacterial proteases, have identified the proteases of chloroplast. They then applied this knowledge based on bacterial proteases was applied to the chloroplast system.

The sequence of proteolysis of D1 can be depicted as follows (Andersson and Aro, 2001) :

$$D1\ protein \xrightarrow[+GTP]{Deg\ P2} 23\ kDa\ N\text{-}term\ fragment \xrightarrow[+ATP,\ +Zn^{2+}]{FtsH} Low\ MW\ fragments$$

There is also a very specific nucleotide requirement for the proteolytic steps and it was shown that the primary proteolytic step is GTP requiring, and the subsequent steps are ATP dependent (Spetea et al., 1999). Haussuhl et al., (2001) have designated the protease responsible for primary cleavage as DegP2 based on their studies on genomic analysis of *Arabidopsis*. The protein is 60 kDa, containing a catalytic triad (serine, histine and aspartic acid) and is located on the outer thylakoid surface. The role of DegP2 has been established when it was added to photoinhibited membranes after removal of endogenous enzymes. $DegP_2$ induces degradation of D1 protein. Therefore, DegP2 protease appears to be the main candidate for the primary D1 protease.

Recently, Chassin et al., (2002) have characterized DegP1 proteins from the thylakoid lumen of *Arabidopsis*. DegP1 was shown to be involved in the degradation of lumenal proteins, plastocyanin, OEC 33 and hence, it is suggested that DegP1 is a general purpose protesase of the thylakoid lumen. Further, it is thought that DegP could cleave the lumenal loops which facilitates translocation and degradation of transmembrane helices by ATP dependent FtsH proteases on the stromal side (Chassin et al. 2002).

It was stated earlier that the primary cleavage occurs on the stromal loop connecting the helices D and E and yeilds N-terminal 23 kDa and C-terminal 10 kDa (Canovas and Barber, 1993; Haussuhl, 2001). The proteolytic cleavage of photodamaged D1 protein is still apparently not understood by way of the precise identity of proteases involved in the process. This, however, is a GTP dependent process and mediated by serine – type protease (Spetea et al., 1999).

Further digestion of primary cleavage products occurs in a second step. An FtsH metalloprotease is involved in secondary proteolysis of D1 protein (Lindahl et al. 2000).

Haussuhl et al., (2001) have isolated from *Arabidopsis* a novel protease DegP2, encoded by DegP2 gene, which is a homologue of the prokaryotic trypsin–type Deg/Htr serine proteases. DegP2 is preferentially associated with the outer surface of thylakoid membrane. The identification and characterization of *Arabidopsis* DegP2 protease which can perform the primary cleavage of photodamaged D1 has been shown. DegP2 is a member of the large family of related Deg/Htr serine proteases. The chloroplast DegP2 is located on chromosone 2 of *Arabidopsis*. It depicts a serine type ATP – independent proteolytic activity (Haussuhl et al., 2001). A common feature of Deg/Htr proteases is the presence of PDZ domain. DegP2 contains 1 PDZ domain located at C terminus.

Haussuhl et al., (2001) have made an indepth investigation of the chloroplast DegP2 protease, whose function is very central to the degradation of D1 protein and the repair cycle under light stress. PDZ domain in DegP2 plays a role in recognition and binding of D-E loop of damaged D1 proteins. The overall sequence of degradation of D1 protein has been shown in a model (Lindahl et al., 2000).

The subsequent breakdown of the primary fragment, 23 kDa occurs through the ATP and Zn, requiring step catalysed by the protease FtsH. It is a 78 kDa with two membrane spanning regions. Lindahl et al., (2000) have conducted a detailed study of the thylakoid FtsH protease purified from *Arabidopsis* and described the turnover of PS II D1 protein. Their study has clearly established that the 23 kDa D1 fragment shows a FtsH dependent protein degradation. This study also suggested that besides FtsH, there may exist other proteases which may also be involved subsequently. *Arabidopsis* FtsH specifically degrades N-terminal 23 kDa. The results described above denote the occurrence of a multistep process of D1 protein degradation. However, the activity of the proteases mentioned is to be precisely identified with either the acceptor side or donor side type of photoinhibition. Detailed study is, therefore, required in this direction.

Recently, Yamamoto (2001) proposed a new pathway for the degradation of photodamaged D1 protein by stromal proteases that indicated photodamaged D1 degrades through membrane bound and stromal proteases. While investigating photoinhibition, it was observed (Yamamoto et al., 1998; Ishikawa et al., 1999) that a cross linking of D1 protein takes place with adjacent polypeptides of PS II RC and core antenna components. It was particularly noticed that cross linking takes the place of D1 protein with D2, α subunit of cytochrome b-559 and CP 43 and the product is further subjected to digestion by a stromal protease. It was also shown that formation of cross-linked products in the donor side photoinhibition is regulated by 33 kDa subunit of OEC on the lumenal side of PS II. More recently, Henmi et al., (2003) have further extended the studies on cross linking of proteins in the donor side photoinhibition. However, this scheme of D1 degradation proposed by Yamamoto (2001) necessitates further study in relation to the existing concepts of D1 protein degradation. Degradation of polypeptides of PS II other than D1 is also known to some extent. The D2 protein is shown to undergo degradation along with D1 protein (Barbato et al., 1992a,b,c). Also, CP 43 undergoes degradation during the donor side photoinhibition (Yamamoto and Akasaka, 1995). Ortega et al., (1999) have demonstrated the degradation of cytochrome b-559 under high light stress.

It is known that a partial disassembly of PS II complex is essential to facilitate the insertion of a new protein copy. The configuration and stabilization of the disassembled PS II core and the precise pathways of removal and replacement of photodamaged D1 are less known (Yokthongwattana et al., 2001). It has been shown that HSP 70 is involved in the damage and repair process of PS II in *Dunaliella salina*. A binding of HSP 70B with the disassembled PS II core complex produces a repaired intermediate in green algae (Yokthongwattana et al., 2001). However, in higher plants, the D1 phosphorylation is considered to prevent premature removal of photodamaged D1 (Baena – Gonzalez, 1999).

That heat shock protein, HSP 70, is involved in the protection of PS II reaction centres during photoinhibition and in the PS II repair process in *Chlamydomonas*, as has been shown earlier by Schroda et al., (1999). The chlorophyll a binding proteins of PS II, CP47 and CP43 can also be degraded. Also, a low molecular weight subunit PsbW has high degradation rate like D1. PsbW protein has been suggested (Shi et al., 2000) as a requirement for dimer formation.

The PS II centres (functional) are in the appressed grana membranes, where photoinactivation and photodamage also takes place, but the repair of PS II and insertion of new D1 copy take place in the stroma exposed thylakoid domains. The new D1 is directly inserted after depletion of the damaged D1 (Zhang et al., 1999). Most of the damaged PS II are actually in the grana, while a few are stroma exposed. After migration of damaged PS II to stroma-exposed domains, proteolytic degradation is initiated. It is understood that since functional PS II centres are located in the grana membrane, the processes of photoinactivation and primary photodamage to PS II take place in the grana only (Adir et al., 1990). The site of repair of PS II and the insertion of the new copy into the thylakoid membrane is the stroma-exposed domain of the thylakoid. It is further known that the insertion of new D1 takes place into the existing PS II after removal of the damaged D1 (Zhang et al., 1999).

2.5B. Resynthesis of D1 Protein and Insertion

The subject of the overall repair process of PS II D1 protein including synthesis of a new copy and the replacement of the damaged D1 protein has been reviewed by Zhang and Aro (2002).

Following the degradation of photodamaged D1 protein, the next step in the repair cycle involves the synthesis of new D1 copy and its insertion into PS II, leading to the restoration of functional PS II. Zhang et al., (2000) have made an exhaustive study of the different aspects of the biogenesis of D1 protein, particularly the regulation of translation elongation, insertion and assembly of PS II. They have also shown that an integration and assembly of D1 protein occurs with other polypeptides of PS II, specifically with D2 protein which apparently acts as a receptor for the D1 protein elongation (Van Wijk, et al., 1997). The insertion and assembly of the chloroplast genome-encoded proteins into the thylakoid membrane is less understood, while the nuclear-encoded thylakoid proteins are better understood, which are inserted post-translationally (Keegstra and Cline, 1999). It is suggested that D1 protein and cytochrome f are inserted into the membrane cotranslationally. A common pathway is also known to exist for the translation of D1, cytochrome f and D2.

The studies of Zhang et al., (2000) using the spinach chloroplast system have shown many crucial steps in the regulation of biogenesis of

D1 protein. It was established that D1 translation elongation was also regulated by redox status besides the translation initiation. The insertion is presumably dependent on transmembrane proton gradient. The destabilization or blockage of several cotranslational and post-translational steps of photosystem II assembly was noticed by the use of thiol reactants. Also, the cross-linking experiments and the results of the thiol reactant experiments suggested an interaction of D1 and D2 proteins. Interaction also exists in the C-terminal processing of the precussor D1 (p D1) and in the reassociation of CP 43 (Zhang et al., 2000). It was shown that carboxyl–terminal extension of precursor D1 protein of PS II is essential for optimal photosynthetic function in *Synechocystis* (Ivelva et al. 2000). Ohnishi and Takahashi (2001) have recently made a study of the multistep process of the repair of the photodamaged PS II in *Chlamydomonas*. It may be important to understand the rates of photodegradation of PS II proteins during the repair process. The rate of degradation of D1/D2 heterodimer is estimated to be t½ = 15 min. which occurred immediately after photoinactivation of oxygen evolution (t½ = 12 min.). However, the photodegradation of CP 43 and CP 47 is comparatively slower (t½ = 25 min.). The repair process involves the replacement of photodamaged D1 and possibly other PS II proteins by the newly-systhesized copies. This is followed by reintegration and reactivation of various cofactors that were lost during replacement. The D1 synthesis occurs in the stroma-exposed thylakoids while PS II functional complexes are located in the grana, as already mentioned earlier. Ohnishi and Takahashi (2001) have studied the role of the small chloroplast-encoded PsbT subunit in the repair process. Their study has established that the chloroplast-encoded synthesis of proteins was not affected in the absence of PsbT but only the post-translational events of repair process are impaired in the absence of PsbT. They have concluded that since PsbT occurs in the vicinity of D1/D2 heterodimer, it may facilitate the binding of the pigments, quinones and OEC to D1 during the assembly into the repaired complex.

It is a well-known fact that the D1 protein is encoded by psb A gene. In chloroplasts, the psbA mRNA ribosome complexes are targeted to thylakoid membrane (Nilson et al. 1999). Only after cleavage and partial degradation of damaged D1, the synthesis of new D1 copy takes place. The D1 is incorporated and assembled into PS II complex (Zhang, 1991). It has been recently shown that translation elongation of D1 has a regulatory role in the assembly of the new D1 copy (Zhang, et al. 2000). Light is essential for PS II repair.

The translation process in the D1 protein synthesis takes place on ribosomes bound to the thylakoid membrane. The translation initiation process involves several steps. Initially, the reducing equivalents produced at PS I activate the chloroplastic protein RB 60 (Kim and Mayfield 1997).

Then this protein assists in the reduction of RB 47 (Danon and Mayfield 1994). Therefore, the multiprotein complex of RB 60 and RB 47 activate the translation initiation process by binding to 5′ – end of psbA mRNA. Sugiura et al., (1998) have suggested that both translation initiation and elongation of D1 are under strict regulation in higher plant systems. It has been established that several pausing intermediates of 17 to 25 kDa are produced by ribosomes during D1 protein elongation (Zhang et al., 1999). The significance of the pausing intermediates may lie in the fact that they may facilitate the stabilization of full length D1 (Zhang and Aro, 2002). Translation elongation of psbA mRNA is dependent on photosynthetic electron transport, particularly on the production of reducing components by PS I (Zhang et al. 2000). The role of the subcomponents of PS II and their assembly has considerable impact on the D1 protein translation. Also, translation of psbA mRNA is inhibited in mutants lacking D2 or CP 47. Cytochrome f also, a chloroplast-encoded protein, is highly relevant for the expression of proteins leading to the correct stoichiometry of the polypeptides at different sites.

Smith and Kohorn, (1994) observed that prevention of translocation of cytochrome f interfered with the translocation of D1 and LHC II. Thus, there exists a common translocation pathway shared by LHC II, D1 and cytochrome-f. The translocation of nuclear-encoded proteins has atleast four distinct routes, which will be described later. Based on the studies in maize and *Arabidopsis* mutants, some of the components in nuclear-encoded protein targeting and translocation are essential for chloroplast-encoded protein targeting and insertion. Hirose and Sugiura, (1996) have developed the chloroplast S 30 translation system, which has prompted the identification of translocational components both for nuclear-coded proteins which are post-translationally translocated as also for the cotranslationally-inserted proteins encoded by chloroplast genome. The whole area of targeting and insertion of proteins across the chloroplast thylakoid membrane is a highly involved area research and is being tackled by a number of workers to understand the mechanisms. Several pathways are recognized (Robinson and Mant, 1997; Robinson et al., 1998; Schcnell, 1998; Keegstra and Cline, 1999; Dalbey and Robinson, 1999). In the Sec A and Signal Recognition Particle (SRP) pathways, the translocation of proteins occurs through nucleotides and soluble factors in the unfolded state of proteins. However, the Δ pH-dependent pathway uses the transmembrane pH gradient for this purpose. A fourth targeting pathway has also emerged, which is a spontaneous insertion pathway (Kogata et al.1999).

It is also not clear regarding the significance of the existence of multiple pathways for translocation of proteins. It might be that the redundancy may provide back up systems, but it is perhaps more deeply related to the problems of specific assembly of different groups of proteins which follow specific pathways. In this connection, it is apparent that chloroplast SRP system is relevant in maintaining membrane proteins in a soluble state during trans-stromal transportation (Keegstra and Cline, 1999). The protein substrate of ΔpH pathway cannot be transported by Sec pathway. The ΔpH pathway is suited for tightly-folded proteins such as OEC 23, but not the Sec pathway (Creighton et al., 1995). The Δ pH pathway requires a membrane-bound protein Hcf 106, which translocates proteins in a tightly-folded form. Hcf 106 was recently isolated from maize by Settles et al. (1997). A Sec dependant pathway, which operates in chloroplasts, has a Sec A homologue and a thylakoid Sec homologue. Zhang et al., (2001) with spinach PS II reaction centre D1 protein, have shown that besides the well-known role of cp SecY in post-translational translocation of nuclear-encoded proteins, its participation in the cotranslational membrane protein insertion is strongly suggested. A chloroplast signal recognition particle (cpSRP) pathway may be important in the D1 protein translocation. An efficient cross linking of cpSRP 54 takes place to the nascent D1 chain (Nilsson et al., 1999). Therefore cpSRP 54 is involved in the targeting process, while another particle, cpSRP 43 is concerned with post-translational targeting. The D1 protein elongation and membrane insertion takes place concomitantly. Kogata et al., (1999) have identified a thylakoid bound cpFtsY in *Arabidopsis thaliana* which is a bacterial homologue of SRP receptor protein. It is considered that cpFtsY participates in the cpSRP dependent protein targeting to thylakoid membranes (Kogata et al., 1999) presumably the LHCP to the thylakoid membrane.

2.5C. Restoration of Functional Assembly of Photosystem II

The functional PS II is maintained by replacement of damaged D1 with the new copy. Clearly, there is a spatial segregation of the site of synthesis and insertion of new D1 copy and the site of occurrence of functional PS II (dimeric). Considerable evidence exists for the role of D2 in the stabilization of D1, followed by a repair of PS II (Zhang and Aro 2002).

Rapid association of newly-synthesized D1 with other PS II core proteins suggests that elongating D1 intermediates possibly start to interact with D1 depleted PS II. It is understood that D1 depleted PS II complexes act as receptors for the elongating D1 since those complexes contain in addition to D2 CP 47, cytochrome b-559 and the small subunit psbI (Morais et al., 1999). Houben et al., (1999), have studied the insertion of leader peptidase (lep), using it as a model protein for the insertion mechanism in pea chloroplast thylakoids. They have concluded that the insertion of the lep occured cotranslationally, but not post-translationally. They also observed that azide, an inhibitor of ATP dependent Sec A activity, decreased the insertion. Their findings have confirmed that using the model system of lep, the insertion process takes place cotranslationally.

Zhang et al., (1999) have made a detailed study of the cotranslational assembly of D1 protein. Their results show that the assembly of new copy of D1 takes place cotranslationally since the nascent D1 chains interact with D2 during the translation elongation itself. Further, CP 47 may be associated with D2 at this stage, while CP 43 assembles with the core complex after the D1 protein translation is completed. Cotranslational incorporation of newly-synthesized D1 into PS II highlights the tight coordination between D1 translation and incorporation into pre-existing core complexes. During D1 elongation and translocation process, TMs of nascent D1 protein exit laterally from translocon and start interacting with other PS II core proteins. These events can be depicted in the model for synthesis, insertion and assembly of D1 into PS II. The small cp SecY translocon plays an active role in directing the D1 nascent chain to their final location in PS II.

C-terminal processing of D1 protein is done by lumenal proteases after termination of translation. Newly-synthesised D1 is cotran aslationally associated, with pre-existing D2 protein during the repair of PS II centre. CP43 is attached to PS II complex (Zhang et al., 2000). The D1 assembly steps take place in stroma-exposed thylakoid, where PS II monomers are assembled. Repair of PS II is not directly dependent on PS II monomers that are probably assembled in stroma thylakoid, to later migrate to grana. The peripheral LHC II complexes are attached and the repaired PS II complexes are found in their dimer form. D1 protein turnover under different light intensities is responsible for the maintainance and control of functional PS II. The structural information regarding the PS II complex has already been clarified earlier. In brief, PS II contains more than 20 proteins and catalyzes water oxidation and reduction of plastoquinone. PS II RC contains D1 and D2 proteins, α and β subunits of cytochrome b-559, PsbI and PsbW proteins (Fotinou and Ghanotakis, 1990). Incidentally, the chloroplast genome encodes all these proteins. Further, the overall redox components of PS II are Chl P680, pheophytin and non-heme iron are bound to D1/D2 heterodimer and quinones. PS II complex also contains the core antennae, CP 43 and CP 47, and oxygen evolving complex, 33, 23 and 17 kDa along with several small proteins whose function is unknown. Presently, the detailed mechanism of the biosynthesis and assembly of PS II is an active area of research. The repair mechanism is to replace the damaged D1 protein with a new copy. The events occurring in the repair process (Zhang and Aro, 2002) are as follows : interconversion and shuffling of PS II dimers and monomers between grana and stroma; partial PS II disassembly D1 degradation and *de novo* synthesis; targeting; membrane insertion;

reassembly of D1 into PS II; release and ligation of cofactors during degradation of D1; and reassembly of PS II.

While initial assembly of D1 into PS II is a cotranslational process, the formation of functional PS II complexes also depends on several post-translational assembly steps. These stages include reassociation of CP 43 and C-terminal processing of precursor D1 protein. CP 43 is a dynamic component in PS II during damage repair cycle of PS II in vivo. C-terminal processing is critical for several post-translational assembly steps of PS II, such as ligation of Mn cluster and reassociation of OEC. During the repair of PS II, the pigments have to be relegated to the newly-synthesized D1 protein. The synthesis of full length D1 protein and its incorporation into PS II are not affected by the inhibitors of chlorophyll and carotenoid synthesis (Depka et al., 1998). Additional factors assisting the PS II complex formation also exist. The heat-shock protien, HSP 70 is one such additional factor. HSP 70 plays a role as a molecular chaperone in the assembly of PS II. Presently, it is known that the *Arabidopsis* genome contains atleast 2500 chloropast targeted proteins, whose functions are yet to be fully established (Abdallah et al. 2000). There may be several more proteins becoming evident from this study that may function in PS II assembly.

Thus, the repair cycle of PS II, which involves a multistep strategy of the replacement of damaged D1 copy and restoration of function of PS II after photoinhibition, is a major aspect of research which requires further clarification in several steps described above (Zhang and Aro, 2002).

2.6. PHOTOINHIBITION OF PS I

While considering the photoinhibition phenomenon, emphasis is invariably placed on damage to photosystem II in view of the fact that PS II is known to be highly sensitive to light stress, as compared to any other chloroplast component. However, it is apparent that photosystem I is also affected by photoinhibition (Hihara and Sonoike, 2001). PS I is a plastocyanin ferridoxin oxido–reductase. The antennae size of PS I has been assumed to be only slightly affected by changes in light environment. Sonoike and Terashima (1994) exhibited that components on the acceptor side of PS I are the site of inhibition. Destruction of iron–sulphur centres, F_A, F_B and F_x was evident (Sonoike 1995). Therefore, the iron–sulphur centres are primary targets of photoinhibition of PS I. P-700 itself is also known to be destroyed under high PFD. The difficulty in determination of PS I activity is one of the reasons for overlooking PS I photoinhibition.

Degradation of one of the R.C. subunits was identified as PsaB protein (Sonoike, 1996). There are Two types of peptide cleavage; one producing a fragment of 51 kDa and the other resulting in 45 kDa and 18 kda. The site of cleavage in the latter is between alanine 500 and valine 501 of PsaB on the loop exposed to the lumenal side between helices 7 and 8 (Sonoike et al. 1997). The degradation of subunits other than PsaB have been reported. Tjus et al., (1999) showed the degradation of PsaA as well as PsaB during photoinhibition of PS I in barley in vivo. There is a general impression that the first site of damage is on the stromal side of PS I. The reduction of oxygen and formation of ROS at the PS I reducing side cause inhibition. Sonoike (1996) has shown that n-propyl gallate, a scavenger of hydroxyl radicals, protected PS I from photoinhibition. Obviously, hydroxyl radical formed by the reaction between reduced iron in the iron suphur centres of PS I and H_2O_2 from oxygen is the cause of inhibition. Photoinhibition of PS I is mainly observed in combination with chilling stress. It is proposed that photoinhibition of PS I is induced at chilling temperatures by a different mechanism from that of photoinhibition of PS II.

The recovery of PS I from photoinhibition is incomplete even after one week, while PS II photoinhibition overcome in a time scale of a few hours only. Thus, photoinhibition of PS I is much more severe for plant survival. It has been shown that a decrease in PS II activity protects PS I from photoinhibition. The PS II activity is down regulated when pH gradient is built across thylakoid membranes. Light energy absorbed by PS I leads to charge separation, which results in oxidation of P 700 and reduction of iron-sulfur centres. Scavenging ROS is an important mechanism for protection. PS I has been found to be susceptible to photoinhibition in an anti-sense mutant of tobacco (Hihara and Sonoike, 2001).

Cyclic electron transfer around PS I is considered to provide protection against photoinhibition. Cyclic electron flow leads to formation of a trans-thylakoid proton gradient, which in turn, leads to down regulation of PS II — and protects PS I — from photoinhibition. Xanthophyll cycle is also associated with PS I light harvesting complex. Xanthophyll cycle pigments bound to PS I could account for 30 to 50 per cent of the total xanthophyll cycle pool (Farber et al. 1997). The PS I photoinhibition is affected by PFD during growth. Under high PFD during growth, PS I is more resistant to photoinhibition (Hihara and Sonoike, 2001). PS I is an important system for response to an ever fluctuating environment. The dynamics of PS I would be an important area of future research.

3
Photoprotection

3.1. INTRODUCTION

Photoprotection in plants generally refers to a multi-component process, evolved to overcome a potential damage arising from absorption of light energy by photosystem II antenna system in excess of consumption during photosynthesis (Horton et al., 1996; Niyogi, 1999; Demmig–Adams and Adams, 1996; Muller and Niyogi, 2001; Mullineaux and Karpinski, 2002; Ort and Baker, 2002). The light energy absorbed by plants in excess of that which can be utilized for photosynthesis is understood to be excess excitation energy (EEE) (Niyogi, 2000; Mullineaux and Karpinski, 2002). Leaves, when exposed to full sunlight, are unable to use all the absorbed light energy to perform photosynthesis (Demmig–Adams and Adams, 1996b). Under relatively optimal environmental conditions, the intercepted light energy by photosynthetic pigments is utilized for carbon assimilation. However, it is thought that photoprotection involves the availability of several alternative electron acceptors under CO_2 limited environment, the detoxification of reactive oxygen molecules and various repair cycles in order to prevent acute photodamage (Ort, 2001). Photoprotection could also be viewed as one of the two components of the overall process of photoinhibition, the other being photodamage. It is understood that photoprotection is dependent on the action of xanthophyll cycle which functions during a shorter time period of less than one hour, while the photodamage results in the inactivation of reaction centre D1 protein of photosystem II and its subsequent repair (Demmig–Adams and Adams, 1992; Long et al., 1994; Thiele et al., 1996; Kitao et al., 2000).

To elaborate the significance of the occurrence of photoprotection in plants, the process is viewed as a balancing measure between the amount of light energy absorbed and utilized and that the inevitable generation of reactive species during photosynthesis is eliminated by antioxidant systems (Niyogi, 1999; Tambussi et al., 2002). When one considers all the climate variables in nature, the critical factor of light intensity becomes apparent, which varies drastically over several orders of magnitude in time and rapidity (Long et al., 1994; Alves et al., 2002). A major plant stress is caused due to high light intensities in excess of that used by photosynthetic process, which obviously leads to an irreversible photodamage to the photosynthetic apparatus (Govindjee, 2002). In relation to light environment, plants should maintain a delicate balance between the maximization of interception of light energy for photosynthesis while mitigating the potential damage due to overexcitation in the chloroplast organelle (Long et al., 1994).

The circumstances under which the absorbed light energy becomes excess are not uniform but rather, highly variable. Plants often encounter an excess of light, when the ratio of Photosynthetic Photon Flux Density (PPFD) to photosynthesis is high (Demmig-Adams and Adams, 1992). Obviously, this condition of excess light refers to a situation where the photosynthetic capacity of the leaf is unable to keep pace with the degree of light interception. Therefore, an increase in the photosynthetic capacity at that stage may partly mitigate the stress caused by high light. Excess light, thus, does not solely depend on the ambient light intensities encountered by the leaf, but also on the environmental conditions that maintain a given capacity for photosynthesis (Ort, 2001; Mullineaux and Karpinski, 2002). Photoprotection process is primarily concerned with the photosystem II function in view of its high susceptibility (Ort, 2001; Ort and Baker, 2002).

Photoprotection against excess light is accomplished by several strategies, which include the preventive type brought through alterations in leaf orientation or chloroplast movements, and the mechanisms that operate before light is actually absorbed by the pigment bed of photosystem II antennae in order to avoid overexcitation (Bjorkman and Demmig – Adams, 1995).

The light avoidance is brought through physiological responses (Anderson et al., 1997, 1998; Park et al., 1997), also viewed as the external factors leading to a reduction in incident light on the leaf (Demmig - Adams and Adams, 1992; Long et al., 1994). This category of plants exhibiting the light avoidance mechanisms are fewer in number while the majority possess other more universal photoprotective mechanisms. Leaves of some plants adjust their orientation parallel to the solar beam to achieve much reduced incidence of sunlight (paraheliotropism) (Kao and Forseth, 1992; Bjorkman and Demmig – Adams, 1995; Koller, 1990, 2000). Paraheliotropic leaf movements exhibited by several legumes avoid the incidence of direct solar beam and, therefore, regulate light harvesting process to balance the capacity of leaf photosynthesis (Kao and Forseth, 1992; Koller, 1990; Sailaja and Das, 1996). Additionally, it is also known that movements of chloroplasts within the leaf have been implicated as a device for decreasing light interception by the chloroplasts. Such movements of chloroplasts are understood to reduce the absorbance of light by as much as 20 per cent (Brugnoli and Bjorkman, 1992; Kagawa and Wada 2002). It has been shown that by the use of mutants of *Arabidopsis* defective in chloroplast avoidance movement, the susceptibility to high light damage was greater in the mutants than the in wild type plants (Kasahara et al., 2002). Kasahara et al., (2002) arrived at the conclusion that the chloroplast avoidance movement results in survival value of plants by decreasing the light absorption, thus offering a protective role against photoinhibition. This observation was also confirmed by a rapid decline in the value of Fv/Fm — an indicator of photoinhibition — in the mutants lacking avoidance movement. Therefore, Kasahara et al., (2002) have clearly demonstrated the role of chloroplast relocation in the photoprotection against high light damage. The interrelations in light harvesting, mechanisms for reduced light absorption, photoprotective strategies and removal (scavenging) of reactive oxygen molecules in higher plants are presented in a schematic diagram in Figure 3.1. The light avoidance mechanisms and also the opposite strategy of diaheliotropism which maximizes the light absorption are discussed in detail in Chapter 4 of this book.

The largest majority of plants lack the ability for adjustment of light absorption and have resorted to strategies for dissipation of the excess photons already absorbed. Such Plants have solved the problem of photoprotection through more universal strategies of removal of excess exitation energy after absorption within the chloroplast. Thus, these strategies constitute internal factors in photoprotection, which are responsible for the removal of excess exitation energy once absorbed. Two principal mechanisms exist for reducing the photo-oxidative damage. One of these involves the thermal dissipation of excess energy in the antennae system through the operation of xanthophyll cycle (Demming – Adams et al., 1996; Gilmore 1997; Gilmore and Govindjee, 1999; Foyer et al., 2000). The second mechanism denotes the oxgen photoreduction to relieve excitation pressure and the later detoxification through antioxidative systems (Asada, 1994, 1999; Demmig–Adams and Adams, 1992; Long et al., 1994; Ort, 2001).

As stated above, a central theme in the overall photoprotection process is the universal mechanism of thermal dissipation of excess energy (non-radiative dissipation of energy as heat), which is a non-photochemical dissipation. This process is a major photoprotective mechanism referred to as non-photochemical quenching (NPQ) (Krause and Weise, 1991; Horton et al., 1994, 1996) since it is measured through the quenching of chlorophyll a flourescence. The photoprotective strategies which often act in concert at different levels in various plants regulate the delicate balance between energy supply and energy consumption (Anderson et al., 1997a). The non-photochemical dissipation is the safest method of removal of excess exitation energy and is regulated by an enzymatic reversible process

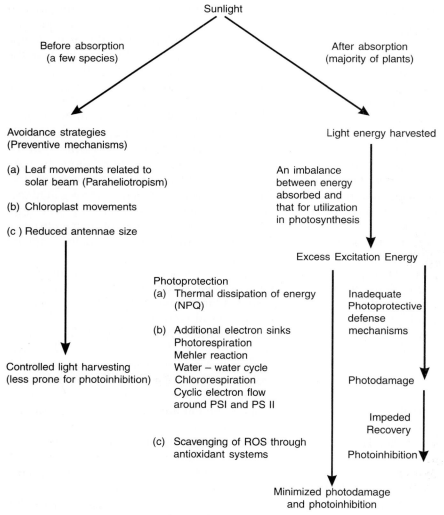

Fig. 3.1. A conceptualized overview of the strategies adopted by plants in the regulation of light absorption and dissipation under natural conditions. Under high irradiance levels plants exhibiting mechanisms of light avoidance minimize the light absorption rather than tolerate high light stress. The more universal phenomenon of excess light absorption involves the built in strategies for mitigation of photoinhibition.

known as the xanthophyll cycle, which depends on the transthylakoid gradient of pH (Δ pH) in the chloroplast. Elucidation of the role of xanthophyll cycle in NPQ is an example of the correlation of biological evidence, driving new biophysical research (Osmond, 1994). Besides the non-photochemical dissipation of excess exitation energy, the photoprotection process also involves the removal or detoxification of reactive oxygen molecules inevitably generated during photosynthesis (Niyogi, 1999; Ort, 2001).

In this context, the antennae size is considered to have an important role in light harvesting and photoinactivation of PS II. However, the relationship of antenna size and light stress is controversial (Park et al. 1997). Cleland and Melis (1987) and Maenpaa et al., (1987) have suggested a dependence of photoinhibition on PS II antennae size, while Tyystjarvi et al., (1994) and Sinclair et al., (1996) opine contrary to this view. Recently, Lokstein et al., (2002) have demonstrated by way of mutant analysis on *Arabidopsis thaliana* that a clear inverse correlation existed between the level of NPQ and the functional PS II antennae size. Teramoto et al., (2002) have shown that excessive light conditions repressed the levels of mRNA of Lhc genes encoding LHC proteins of PS II in *Chlamydomonas*. This finding implies that the size of antennae is expected to decrease under high light intensities. Teramoto et al., (2002) in their study, have demonstrated the presence of two alternative mechanisms in the repression of Lhc genes under high light. One of the these pathways is independent of photosynthetic reaction centres and electron transport. This redox-independent mechanism is presumably involved under stressful light situations, minimizing excitation pressure in order to facilitate the repair of damaged PS II.

3.2. THERMAL DISSIPATION OF EXCITATION ENERGY

In nature, a majority of plants absorb much higher levels of solar energy over that actually required for photosynthesis, thus, resulting in excess light (Long et al., 1994). At low light intensities (around 100 µ mol, m^{-2} s^{-1}), nearly 80 per cent of absorbed energy can be utilized for photosynthesis. As light intensity increases, the amount of light utilized becomes progressively lower. At half of full sunlight (1000 µ mol, m^{-2} s^{-1}), less than 25 per cent of light absorbed is used, while at full sunlight (2000 µ mol, m^{-2} s^{-1}), the utilization is only 10 per cent (Long et al., 1994). Therefore, it is apparent that plants have to adjust their light harvesting capacity through several strategies, in view of the fact that the sunlight varies over several orders of magnitude during a wide range of time scale (Muller et al., 2001).

There exist several avoidance mechanisms — both short term such as chloroplast and leaf movements and long term such as antennae size adjustments — leading to a reduction in the harvesting capacity (Demming - Adams and Adams, 1992; Long et al., 1994; Park et al., 1997; Horton et al. 1996). Despite these measures, plants still absorb light which becomes excess. Therefore, there is a need for existence of a protective mechanism against damaging excess light. The non-radiative (thermal) dissipation of energy in the predominant process for the short-term regulation of light energy harvesting through safe dissipation as heat (Krause et al., 1983, Crofts and Yerkes, 1994, Demmig – Adams et al., 1996). It has been observed that thermal dissipation can dissipate as much as 50-75 per cent of light energy already absorbed (Demmig – Adams et al., 1996; Peterson and Havir, 2001, review by Shikanai et al., 2002). The protection of reaction centres of PS II from overexcitation is actually achieved by locating NPQ in the antennae system (Melis, 1999). Thus, non-photochemical energy dissipation is a potential and major photoprotective process leading to deactivation of singlet excited chlorophylls by thermal dissipation (Demmig – Adams and Adams, 2000). A reduction in delivery of energy from the antennae to PS II RC takes place by non-photochemical dissipation. It also minimizes the generation of singlet oxygen in LHC II and PS II RC (Demmig – Adams and Adams, 2000; Havaux and Kloppstech, 2001).

The light-harvesting antennae are highly efficient in the capture of light energy but it is also essential that plants switch over from harvesting to energy dissipation under excess light (Horton et al., 1996). Accordingly, it is envisaged that without a delicate switching ability between light harvesting and dissipation, the efficiency of chloroplasts would be reduced. This becomes particularly evident in a rapidly-fluctuating natural light environment either by inadequate light absorption or by harmful overexcitation (Andersson et al. 2001). Regulation of light harvesting is an essential attribute of plant photosynthesis to balance the absorption and utilization of light energy and, accordingly, avoid photo-oxidative damage. Light harvesting complexes associated with the reaction centres are responsible for absorption of sunlight. Light absorption results in singlet excitation of Chl a molecule which can return to the ground state *via* the number of different pathways. The pathways of de-excitation take place through: (a) total fluorescence; (b) energy-utilizing photochemical reactions (qP); (c) thermal dissipation (NPQ); or (d) by singlet chlorophyll being converted to long-lived triplet state. The triplet state is also responsible for generating singlet oxygen by interaction with triplet oxygen (Niyogi, 2000; Demmig – Adams and Adams, 2000; Muller et al., 2001; Govindjee, 2002).

The photochemical utilization of energy leads to quenching of fluorescence and under saturation of electron flow, the fluorescence yield increases (Van Kooten and Snel 1990; Schreiber et al., 1998; Lazar, 1999). The photochemical quenching is referred to as qP. When qP is zero (light saturation for electron transport), still considerable chlorophyll fluorescence occurs which is, therefore, contributed by non-photochemical quenching, qN (Schreiber et al., 1986) or NPQ (Bilger and Bjorkman,

1994). Light stress causes dissipation of excess energy through non-radiative decay, which is a major proportion of qN comprising qE, while a minor component is referred to as qI.

As mentioned earlier, the processes of non-radiative harmless dissipation of excitation energy that quench chlorophyll flourescence are termed as non-photochemical quenching, NPQ or qN. (Krause et al., 1983; Demmig-Adams et al., 1996; Maxwell and Johnson, 2000). It is known that increases in non-photochemical quenching take place at light intensities below that of light saturation of photosynthesis (Ort and Baker, 2002). This was also evident in a recent study by Lawson et al., (2002) who have demonstrated that NPQ increased upto a certain level of light intensity, while perceptible decreases in photochemical quenching have determined the efficiency of photosynthesis at higher light levels. Various physiological processes that are responsible for the total compliment of NPQ are diverse and comprise two categories in higher plants and a specific type restricted to algae (qT) (Demmig – Adams, 1990; Horton, et al., 1994; Horton et al., 1996; Owens, 1996; Muller et al., 2001; Garab et al., 2002). The major and most rapidly relaxing ΔpH dependent and xanthophyll cycle controlled component is termed qE. PsbS protein is also involved in this type (Li et al., 2000). The other constituent of NPQ is a slow relaxing and less understood mechanism called qI (Sehreiber et al., 1998). This process is considered to be primarily related to photoinhibition (Horton et al., 1996). Here, qI is induced substantially under conditions of high light treatment. Recently, it has been demonstrated that a significant induction of qI is associated with the structural changes in LHC II, leading to the monomerization of complexes from their usual trimeric configuration (Garab et al., 2002).

The energization of thylakoid membrane and actually the trans thylakoid ΔpH is responsible for qE. A correlation between sustained zeaxanthin–dependent energy dissipation and thylakoid protein phosphorylation was observed by Ebbert et al. (2001). qI, the slowly-relaxing component of NPQ, is ascribed to processes including photoinhibition and PS II damage as also the consequence of zeaxanthin formation (Horton et al., 1994, 1996). qE is understood to be a feedback control mechanism under excessive illumination, sensed through the magnitude of trans-thylakoid ΔpH (Horton et al., 1994).

As already mentioned above, qE, the major component of qN is obviously the dominant process that regulates the light energy utilization in PS II (Horton et al. 1994, 1996). It is understood that the magitude of qE is regulated by the size of the trans-thylakoid ΔpH. Thermal dissipation is regulated by lumenal pH for maintenance of maximal photosynthesis (Munekage et al., 2001). Munekage et al., (2001) have shown that the mutation in the Rieske subunit of the mutant pgr1 (proton gradient regulation) of *Arabidopsis* results in lowering of electron in transport through cytochrome b6f complex. This, in turn, did not reduce the lumenal pH below pH 6.0. Therefore, in the *Arabidopsis* mutant pgr1, there was little build up of Δ pH, insufficient to cause thermal dissipation of energy. Good correlations between increases in qE with decreased quantum yield of photochemistry of PS II RC demonstrate the principal role of qE in the regulation of light energy utilization (Weis and Berry 1987). The presence of zeaxanthin is generally necessary for maximal qE in vivo, but by itself, it is not adequate for the process. Therefore, the low Δ pH has an additional role in qE besides activation of xanthophyll cycle. Presently, the understanding is that the development of NPQ involves both pH and xanthophylls while zeaxanthin may also have other roles of protection (Govindjee and Sewfferheld 2002). The proteins in LHC responsible for qE have been identified through obtaining mutants in *Arabidopsis*. One mutant, npq 4-1, was shown to lack qE and also the confirmational change. PsbS protein is essential for qE (Li et al. 2000). This protein belongs to LHC super family but possesses four transmembrane helices instead of three.

It is yet to be established whether the involvement of xanthophylls is direct or indirect in NPQ. Both violaxanthin (V) and zeaxanthin (Z) could potentially accept energy from singlet chlorophyll. qE and xanthophylls are important for photoprotection and the photoprotective effect of qE is due to decreased production of singlet oxygen.

Two different mechanisms are proposed for the role of zeaxanthin in NPQ (Frank et al., 2000). Of these, one involves a direct quenching. This process occurs through a downhill energy transfer from Chl a to zeaxanthin subsequent to a structural change in the pigment protein complex. The concept of direct quenching is based on the determination of energies of the lowest singlet states of xanthophylls. Excited state energy levels suggest that the lowest singlet state of zeaxanthin and antheraxanthin (A) can accept excitation energy directly from singlet Chlorophyll (Owens, 1994; Frank et al., 1994; Frank et al., 2000). Subsequently, excited xanthophylls return to the ground state by non-radiative heat dissipation. It has been shown that the energy of S_1 state of xanthophylls decreases with increasing number of conjugated double bonds.

Differences in the extent of conjugated double bond systems (CDB) of the xanthophyll cycle carotenoids (Vn = 9 CDB, An = 10 and Zn = 11) directly affect their S_1 energies. S_1 state life times of t = 23.9 ps for V, t = 14.4. ps for A and t = 9 ps for Z and these values were used to calculate the energies of S_1 levels. They were estimated to be 15,290 cm^{-1}, 14,720 cm^{-1} and 14,170 cm^{-1} for V,A and Z respectively (Young et al., 1997; Horton et al., 1999).

That differences in the S_1 energies of V and Z may account for the operation of XC was proposed (Owens et al. 1992). The energy of the lowest exited state of Chl a in LHC of PS II has been estimated in the range of 14700 to 15000 cm^{-1}. The value of 14,700 for Chl a is lower than S_1 state of V but higher than that of Z. This would suggest that it is energetically possible for S_1 of Z to quench Chl flourescence. In constrast, S_1 level of V would only permit the same as a light- harvesting pigment. Therefore, the S_1 state of highly conjugated zeaxanthin lies below that of chlorophyll so that it is unable to function in light harvesting (DeCoster et al., 1992). Accordingly, zeaxanthin is supposed to act as a direct quencher of singlet chlorophyll.

S_1 state of violaxanthin is considered to be above that of Chl a and accordingly participates in light harvesting. Zeaxanthin with increased conjugation length of eleven double donds has its S_1 state below that of Chl a. The detailed spectroscopic and kinetic investigations have revealed that S_1 states of xanthophylls are lower than the previously held view and, therefore, they are able to quench exited singlet chlorophyll states (Frank et al. 2000).

A second mechanism of quenching is thought to be an indirect method involving the binding of xanthophylls to proteins. The structural features of xanthophylls exert a control over the arrangement (organization) of pigments within the antennae complexes. This leads to quenching of fluorescence (Young et al., 1997). Using isolated complexes, it has been established that zeaxanthin promotes the aggregation of antenna complexes, resulting in quenching of chlorophyll fluorescence (Ruban et al. 1998). Further Wentworth et al., (2000) by using isolated LHC II complexes from spinach, concluded that a zeaxanthin-dependant alteration in chlorophyll environment is responsible for the quenching, which is not actually caused by aggregation of the complexes. The strongest evidence supporting indirect mechanisms comes from isolated LHC II. The aggregation or confirmational transitions is the basis for qE.

The most important effect of XC is on the aggregation of LHC II b, CP 29 and CP 26. Xanthophylls play an important role in determining the structure and function of LHC II. De-epoxidation allows the switch between two different states of organization. In the presence of V, complexes are resistant to Δ pH-dependant quenching and therefore optimized for energy utilization. The key feature of the regulation of XC is the size of the pool of VA and Z, which is higher in plants in high light. It has been estimated that PS II in spinach has a binding site capacity of 18 to 21 xanthophyll cycle carotenoids. Caffarri et al., (2001) have studied a xanthophyll–binding site called V1 in the major LHC of PS II which was found to be distinct from the three tightly-binding sites L1, L2 and N1. It has been suggested (Caffari et al., 2001) that probably the LHC II V1 site serves as a source of readily-available violaxanthin for de-epoxidation. Hurry et al., (1997) have shown that light-dependent conversion of violaxanthin to zeaxanthin is not required for formation of

NPQ and zeaxanthin was present in the darkness. Development of NPQ, however, required generation of pH gradient in plants with large pools of zeaxanthin. This suggests that within the LHC antenna of PS II, only a few molecules of zeaxanthin are sufficient for NPQ formation. Large pools of xanthophyll or higher rates of conversion of violaxanthin to zeaxanthin are not necessary for enhancemnet of NPQ. Recently, Bukhov et al., (2001b) have also shown that only a few molecules of zeaxanthin per reaction centre of PS II are effective enough to dissipate light energy in a moss.

Havaux and Niyogi (1999) have clearly demonstrated using a mutant npq 1 of *Arabidopsis thaliana* that violaxanthin cycle preferentially protects thylakoid membrane lipids from photo-oxidative damage. This protective action of xanthophyll cycle is different from that of quenching and singlet excited chlorophylls through NPQ. In mutant npq 1, zeaxanthin is absent due to lack of functional VDE. This absence of zeaxanthin presumably impairs α-tocopherol recycling and, accordingly, vitamin E level is reduced, which in turn enhances lipid peroxidation (Havaux and Niyogi, 1999), thus suggesting that xanthopyll cycle has more than one role. These roles include both the quenching and prevention of lipid peroxidation.

3.2A. Site of quenching

Quenching through qE can occur either in the antennae system of PS II or in the reaction centre after the excitation has been trapped. The quenching in a reaction centre is perhaps related to charge recombination between P680 + and Q_A^-. Energy trapped by PS II is dissipated as heat (Krieger, et al., 1992). However, this mechanism of quenching in a reaction centre needs further confirmation. In this context, recently, Peterson and Havir (2001) have determined the key process of NPQ in normal and mutant leaves of *Arabidopsis thaliana*. They observed that in wild type (normal) leaves in the magnitude of photochemistry in open PS II centers, there was only an insignificant decrease with increase in H^+ dependent thermal dissipation (qE). Accordingly, they concluded that NPQ was primarily localized in the antenna complex. However, in mutant npq 4-9 (with substituted amino acid in PsbS gene and reduced NPQ), there was a substantial decrease in the photochemistry with increasing qE, therefore, suggesting that NPQ is localized in PS II RC. Based on these observations, it is suggested that RC quenching occurs when antennae quenching is inhibited.

Detailed studies of light energy dissipation were made in the leaves of *Spinacea oleracea* and *Arabidopsis thaliana* and in a moss using chlorophyll fluorescence (Bukhov et al. 2001a). The experiments of Bukhov et al., (2001a) have suggested that the antennae quenching was the predominant dissipation mechanism in the moss, while RC quenching was for spinach and *Arabidopsis*. They further observed that both types of quenching depended on thylakoid protonation, while only antennae quenching depended on zeaxanthin. The slow and reversible quenching, referred to as photoinhibitory quenching, qI, persisted after deprotonation and presumably originated in RC.

It was earlier found (Rees et al., 1990) that in pea leaves, qE originates from the antennae. In contrast to these findings, Bukhov et al., (2001a) found that RC quenching is more affective than antennae quenching in spinach leaves. The assignment of qI quenching to RC, which was suggested by Richter et al., (1999) was confirmed by the studies of Bukhov et al. (2001). RC quenching is also suggested to be the mode in the zeaxanthin deficient mutant of *Arabidopsis*. Photoprotection is generally ascribed to events in the antenna complexes of PS II which involves the de-epoxidation state of xanthophyll pigments and the trans-thylakoid Δ pH (Forster et al., 2001). Presently, it is believed that PS II antenna is the actual site of NPQ for which a general consensus is steadily emerging (Govindjee, 2002).

However, the whole field of the study of localization of NPQ within the PS II complex still requires further study in a larger number of plants and appears to be a fertile field for future research.

It is pertinent to recapitulate that the PS II antenna is composed of two different protein complexes, designated as core and peripheral. The core inner complex is a CP 43 and CP 47 Chl a and carotene binding protein encoded by plastid genome. The peripheral (outer) complex comprises LHC II (major) and minor pigment proteins comprising CP 29, CP 26 and CP 24 and are both encoded by nuclear genome (Lhcb4, Lhcb5, Lhcb6). These minor complexes are mono-meric, while the major LHC II b comprises trimeric polypeptides.

LHC II b comprises the main trimeric poplypeptides –28, 27 and –25 kDa encoded by genes Lhcb 1, Lhcb2, and Lhcb 3. The minor LHC II a (29), LHC IIc (26) and LHC IId (24) bind only 5 per cent of PS II chl and they bind the PS II core to LHC IIb (Horton et al. 1996).

3.2B. Mechanism of qE

The process of qE could take place through the following steps: light-induced H^+ transfer and binding of H+ to LHC II polypeptides; and conversion of violaxanthin to zeaxanthin, resulting in a quenching complex bound to LHC II. The quenching complex could either be a chlorophyll zeaxanthin or Chl – Chl. A direct energy transfer to zeaxanthin would occur in the former situation; in the later case the xanthophyll is considered to modulate the formation of complex (Horton et al., 1996). The two components — protein protonation and zeaxanthin availability — are responsible for the dissipation of energy. Hartel et al., (1996) have conducted kinetic studies of the xanthophyll cycle in barley leaves and investigated the influence of antennae size in relation to NPQ. It was found that the capacity to develop qN was reduced step-wise with the antennae size. The study provided evidence for structural changes in LHC proteins that form the basis for development of NPQ. The fact that protons play an important role in NPQ of chlorophyll a fluorescence has been confirmed in a study using a mutant of pH gradient in *Arabidopsis thaliana* (Govindjee and Spilotro, 2002).

As described earlier, the relative effects of the different carotenoids in fluorescence quenching have been studied in an in vitro system by Ruban et al. (1998), who suggested that quenching by way of xanthophyll cycle carotenoids is through an indirect structural effect and not by direct quenching. For direct quenching, the key feature is reduction in S_1 energy level of Z arising from the shortening of carbon double bonds (CDB) from 11 to 9. This can be obtained by the removal of two epoxide groups from the carotenoid. Ruban and Horton, (1999) have made an analysis of the kinetics of NPQ in spinach at different levels of organization: leaves, chloroplasts and isolated LHCs. They arrived at the conclusion that the principal effect of low pH is due to protonation induced confirmational changes in the components of PS II antenna. Therefore, this view supports the requirement of both protonation and zeaxanthin availability for the induction of confirmational changes. De-epoxidation brings about significant changes in structures and hence properties of carotenoids. It is understood that when conjugated chain length is increased in carotenoids from 9 to 11 double bonds, it not only affects their S_1 energies but also modifies the size and shape of the molecule. The direct quenching of chlorophyll fluocerscence occurs through singlet-singlet energy transfer to zeaxanthin while violaxanthin transfers energy to chlorophyll, which can be explained by a molecular gear – shift model (Horton et al., 1999). The molecular gear shift was proposed to describe a model for interconversion of V into Z, which serves to alter their S-energies. S_1 energies and life times of XC carotenoids have not been determined in vivo and there is no direct evidence to show that singlet–singlet energy transfer for Chl to carotenoid occurs in vivo. The molecular gear shift scheme described above may be a possible route of deactivation.

The precise molecular mechanism of qE is not yet elucidated but the following points are worthy of note:

1. Quenching is caused by Δ pH through thylakoid lumen acidification (Noctor and Horton, 1990).
2. qE is associated with a conformational change (Ruban et al., 1993; Bilger and Bjorkman; 1994).
3. qE occurs in LHC of PS II. LHC II comprises products of six Lhc b genes (Lhc b 1-6) assembled into 4 complexes known as LHC

IIa, LHC IIb, LHC IIc and LHC IId. LHC IIb is trimeric and binds 60 per cent of PsII Chl. LHC IIa, LHC IIc and LHC IId are monomeric, widely known as CP 29, CP 26 and CP 24, respectively. qE occurs in one or more of these complexes (Horton, 1994, 1996), but qE is reduced when Lhc polypeptides are absent (Jahns and Krause, 1994). qE is eliminated when LHC II associated zeaxanthin and lutein are missing in *Chlamydomonas* double mutant. The precise location of thermal dissipation within the LHC II and the actual polypeptides responsible for NPQ are still a matter of considerable debate. An earlier work has implicated both the LHC II trimers (Horton et al., 1991) and monomers (Bassi et al., 1993; Crofts and Yerkes 1994) for the location of thermal dissipation. Using a barley mutant, it was shown that the trimeric LHC IIb is perhaps not the site for NPQ (Gilmore et al. 1996). Contrary to this finding, Chow et al., (2000) studying plants transferred to continuous light from intermittent growth light suggested that LHC IIb has a definite role in NPQ as there was a positive correlation in their levels. However, recently, Elrad et al., (2002) have made a detailed investigation of the thermal dissipation process using npq-5 mutant of *Chlamydomonas*. The mutant is deficient in the gene $Lhcbm_1$ that encodes a light-harvesting polypeptide located in the trimers of PS II antenna. This study (Elrad et al., 2002) has established a role for LHC II trimer in the thermal dissipation. In this connection, it is known that in *Arabidopsis* plants the PsbS component of PS II — whose precise location is not established — has been shown to be the site of NPQ (Li et al. 2002). Since PsbS polypeptide is not located in *Chlamydomnas*, it is presently not clear whether the situation is different in higher plants. Therefore, this aspect of research related to the location of NPQ requires further clarification. Also, recently Grasses et al., (2002) have examined the role of ∆pH – dependent dissipation of excitation energy in mitigating the damage caused by high light stress in *Arabidopsis*. Grasses et al., (2002) have, in fact, concluded that energy dissipation in PS II is not primarily required for affording protection of PS II in high light but is presumably involved in reducing electron pressure in the photosynthetic electron transport chain under light saturation. Therefore, detailed research is needed in understanding the roles of qE mechanism in the overall photoprotection.

A strong correlation between the levels of Z and qE indicates that conversion of violaxanthin to zeaxanthin controls qE. ∆ pH and Z together control the induction of quenching. Evidence exists that quenching of excitation energy takes place in LHC II rather than in PS II RC or core antennae (Ruban and Horton, 1994; Horton et al., 1999). It has been suggested that minor complexes CP 29 and CP 26 may contain the quenching sites. However, in a recent work (Andersson et al., 2001) using *Arabidopsis* transgenic lines, it has been shown that CP 29 and CP 26 are unlikely to be the sites for NPQ (Shikanai et al., 2002). In this context, the studies of Li et al., (2000) have proved shown that the PsbS gene product as the site of NPQ. Also, cold acclimation involves increased NPQ capacity by adjusting PsbS abundance. However, Rorat et al., (2001) while investigating the chilling stress in *Solanum*, have observed that neither PsbS nor NPQ were actually involved in low temperature acclimation of *Solanum*. They have, on the contrary, found that the levels of two chloroplastic drought-induced stress proteins (CDSP) were higher in cold–resistant species. Recently, however, Li et al., (2002) have investigated the overexpression of PsbS through generating plants of *Arabidopsis* with a twofold increase of qE capacity. It is already known the qE development requires a low thylakoid lumen pH, xanthophyll de-epoxidation and the PsbS, a photosystem II protein. Li et al., (2002) have clearly demonstrated that in *Arabidopsis*, the level of PsbS is well correlated with qE capacity. It was also shown that increased qE was associated with excitation pressure on PS II and that a positive correlation existed with qE capacity and resistance to photoinhibition. From these studies, it was suggested that in future an engineered qE capacity by overexpression of PsbS might produce greater resistance to light stress (Li et al. 2002).

In essence, the induction of qE is brought out in the following manner.

The processes of de-epoxidation of V → Z via A and subsequent epoxidation alters the structure of these pigments. The structural effects are described below :

Changes in carotenoid structure (as already noted earlier) directly result from: (a) lengthening (de-epoxidation); or (b) shortening (epoxidation) of the conjugated double bonded system (CDB), which ranges from 9 to 11 carbon – carbon double bonds in V and Z, respectively.

The implications of such changes are twofold.

i. The extent of CDB affects the energies and life times of their excited states.
ii. The extension of CDB into the β-end group of carotenoids introduces configurational changes.
iii. The physio-chemical properties of a carotenoid moleules are linked to its structure and, therefore, it is important for the biological role of xanthophyll cycle carotenoids. It requires less energy to aggregate carotenoids with two β - end group such as β carotene and Z than carotenoids that posses either e – end groups or epoxides (V).

3.3. XANTHOPHYLL CYCLE

In view of the significance of xanthophyll cycle and its close connection with the NPQ, it is pertinent to consider the pathway of cycle and its characteristics in some more detail. The principal photoprotective function of the xanthophyll cycle is to increase heat dissipation. It is understood that the non-photochemical energy dissipation is predominantly xanthophyll cycle dependent (Gilmore, 2001). It is also established (Niyogi et al., 1998) by studies on mutants of *Arabidopsis* and *Chlamydomonas* that the role of xanthophyll cycle in NPQ has been conserved.

Xanthophyll cycle is defined as a light dependent, reversible conversion of xanthophylls, violaxanthin to zeaxanthin via antheraxanthin (Jahns, 1995). Recent studies from several laboratories have established the importance of this cycle in the regulation of light harvesting, protective function against overexcitation and recovery from photoinhibiton. The risk of photodamage is considered to be reduced by zeaxanthin.

The cyclical interconversion of carotenoids, violaxanthin, antheraxanthin and zeaxanthin takes place in plants and green algae during the xanthophyll cycle. The cycle involves de-epoxidation and re-epoxidation reactions that interconvert violaxanthin (V) antheraxanthin (A) and zeaxanthin (Z) (Yamamoto, et al. 1962). The main focus of xanthophyll cycle is its role in protecting plants against the potentially damaging effects of excess light — when irradiance is greater than that which can be used in photosynthesis. Excess irradiance is species specific and depends on several factors. The de-epoxidized pigments, Z and A, enable excess energy absorbed by chlorophyll to be harmlessly dissipated as heat (Hartel et al., 1996; Yamamoto et al., 1999).

The unutilized and excess excitation energy might cause decay, leading to production of toxic reactive intermediates including singlet oxygen arising from triplet chlorophyll. Such a contingency is avoided by the existence of an alternate pathway involving elimination of the excess excitation energy by harmless dissipation in the form of heat, through a direct de-excitation of singlet chlorophylls (Demning – Adams et al., 1999). To prevent such de-excitation through undesirable pathways, the excess exitation energy is dissipated through harmless heat.

A view is held that two factors — an association of zeaxanthin and antheraxanthin with proteins of light harvesting pigment bed of PS II and the protonation of the LHC polypeptides — catalyse the dissipation of excess excitation energy (Yamamoto et al. 1999). Considerable work has been carried out by various workers in order to understand the interrelations among thylakoid acidification, xanthophyll cycle de-epoxidation and energy dissipation (Horton et al., 1996; Gilmore, 1997). The thermal dissipation of energy is induced by thylakoid acidification serving as a signal for excess light, while the actual energy dissipation depends on the presence of zeaxanthin and antheraxanthin (Gilmore et al., 1998). The process of thermal energy dissipation can be viewed as affording photoprotection while allowing plants to keep the investment in photosynthetic capacity to a level required to support the growth rate under genetic constraints and environmental conditions.

3.3A. Mechanism

The cycle was discovered by Sapozhnikov et al., (1957), following which the details were elaborated by research groups of Yamamoto and of Hager (Yamamoto, et al., 1962; Hager, 1969; Yamamoto, 1979; reviews by Yamamoto et al., 1999; Deming - Adams et al., 1999; Eskling, et al., 2001). The actual mechanism of events in the cyclical process is depicted in Figure 3.2.

The principal reactions are catalysed by two enzymes — Violaxanthin De-epoxidase (VDE) and Zeaxanthin Epoxidase (ZE). The enzyme, VDE is located in the thylakoid lumen and its action is dependent on the pH of the lumen. It is a soluble enzyme at high pH but due to the generation of proton gradient under irradiation by electron transport, the pH of the lumen drops, when the enzyme gets attached to the inner thylakoid membrane. VDE is specific for xanthophylls that have 3-Hydroxy-5, 6-epoxy group in a 3S, 5R, 6S configuration and are all trans in the polyene chain. Grotz et al., (1999) have suggested that epoxy–xanthophylls available for VDE occur as rod–like and transconfigurated pigments within the lipid bilayer of thylakoids. There is a strict ascobrate requirement (Bratt et al., 1995) alongwith the need for major thylakoid lipid, mono galactosyl diacyl glycerol (MGDG). Conversion of violaxanthin to zeaxanthin takes place through the formation of intermediate antheraxanthin. Rockholm and Yamamoto (1996) showed a direct relationship between thylakoid lipids and VDE. Their studies clearly demonstrate that MGDG is four times more effective in precipitating VDE in comparison with DGDG and is nearly 38 times more powerful than other lipids. Havir et al., (1997) have purified VDE from spinach and the properties of the enzyme were studied in detail. The studies also emphasized the requirement of 50 M of ascorbate and a threefold stimulation by MGDG at a pH optimum of 5.2. The enzyme was purified with an yield of 1.4 per cent and 190-fold purification (Havir et al., 1997) and the sensitivity of the purified enzyme to sulphydryl compound DTT was demonstrated using violaxanthin as substrate. The enzyme activity was inhibited by 50 per cent, a finding which was in agreement with that shown for the DTT inhibition of the enzyme by other workers (Yamamoto and Kamite, 1972; Neubauer, 1993; Gilmore and Yamamoto 1993).

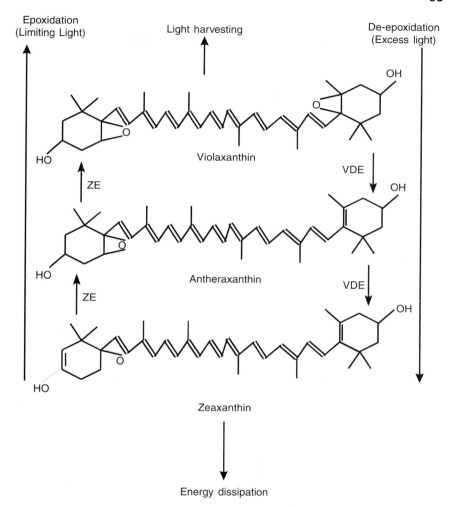

Fig. 3.2. The reversible xanthophyll cycle and the structures of the pigments. The carotenoids have dual function of light harvesting under low light conditions and of energy dissipation in excess light. The de-epoxidation increases the conjugation length while in the reverse, the epoxidation decreases the number of double bonds.

The reaction sequence is as follows :

$$\text{Violaxanthin} \xrightarrow{\text{VDE (43 kDa)}} \text{Antheraxanthin} \xrightarrow{\text{VDE (43 kDa)}} \text{Zeaxanthin}$$

The xanthophyll cycle the is chloroplastic, asymmetrically organized across the thylakoid membrane with de-epoxidation on the lumenal and the reverse reaction, and epoxidation on the stromal sides of the membrane.

The reversible reactions of de-epoxidation and epoxidation catalysed by the enzymes VDE and ZE have pH optima of 5 and 7.5, respectively (Yamamoto et al. 1999). While VDE is ascorbate dependent, ZE is NADPH and O_2 dependant. As already mentioned, VDE is responsible for the reaction sequence V ———> A ———> Z, a step-wise removal of 5,6 epoxide, perhaps by reductive dehydration.

The low pH of lumen necessary for de-epoxidation is the photosynthetic proton pump. The requirement of ascorbate for the reaction of VDE suggests that the de-epoxidation mechanism involves a reduction followed by dehydration. The enzyme VDE, which has been purified from spinach (Kuwabara et al., 1999) and lettuce (Rockholm and Yamamoto, 1996), has been shown to be a 43 kDa protein. V is a di-epoxide with 9 conjugated double bonds. Its S_1 state is said to be well above that of Chl a, which would, therefore, allow transfer of energy fron V to Chl. Conversion of V to Z removes the epoxides and increases the conjugation to 11 double bonds. This conversion lowers the energy of zeaxanthin S_1 state, which would enable chlorophyll a to transfer energy to Z. Accordingly, the function of V is light harvesting, while that of Z is energy dissipation. The de-epoxidation is brought about by increased conjugation length and the epoxidation by reduction in the conjugation length. The energy levels also decrease, with an increase in the number of double bonds.

VDE acts in vivo on V, in the lipid phase of thylakoid. Therefore, V, associated with LHC II, is in equilibrium with the lipid phase. There is also work done on cis-trans isomerization in regard to V. It is said that from cis to all trans isomerization is a requirement for the availability. The reversible epoxidation sequence V ← A ← Z occurs in darkness and is slower than the forward de-epoxidation. ZE has not been isolated directly from plants. ZE is a mono oxygenase using molecular oxygen and NADPH, while FAD is a cofactor (Buch et al., 1995). FAD occurs in darkness and is stimulated by weak light. NADPH and ferredoxin are required for the ZE activity. The activity is carried out in the major thylakoid lipids MGDG and BGDG. Therefore, epoxidase may also require all-transconfiguration, as in the case of de-epoxidase. Ascorbate, an abundant antioxidant in plant chloroplasts, is used as a reductant for the conversion of violaxanthin through the action of VDE. The functions of both pH and ascorbate concentration was studied by Bratt et al. (1995). The uncharged acid form of ascorbate would be the only form able to penetrate the membrane. The details of regulation of VDE by ascorbate have been recently reviewed by Eskling et al. (2001).

Eskling and Akerlund (1998) have studied the changes occurring in the levels of VDE, xanthophylls and acorbate in spinach during transfer from low to high light. Z, a carotenoid in the xanthophyll cycle, plays a crucial role in preventing photodestruction. Eskling and Akerlund (1998) have examined the levels of these constituents in plants when transferred from low light (100 – 250 μ moles/m^{-2} S^{-1}) to high light intensity (950 μ moles m^{-2} S^{-1}). VDE decreased by nearly 30 per cent, Xanthophyll cycle pigments doubled and the rate of V → Z conversion was also doubled. Lutein and neoxanthin also increased. Eskling and Akarlund (1998) arrived at the conclusion that the increase in xanthophyll cycle pigments and ascorbate only partially explains the increased rate of conversion of V → Z, but the most probable explanation of faster conversion is due to the increased acessibility of V in the membrane. The precisc mechanism for dissipation of excess energy by xanthophylls is controversial. The dissipation is measured as non-photochemical flourescence quenching (NPQ). It was pointed out in the earlier section that there are two hypotheses for the operation of xanthophyll cycle. One of them envisages a direct role for the de-epoxidized xanthophylls in producing quenching complexes with light harvesting (Gilmore, 1997) while another proposes an indirect role through the de-epoxidation, leading to enhanced production of aggregated light harvesting complexes which are highly quenched (Young et al., 1997; Horton et al., 1996). However, there is a general consensus that xanthophyll cycle provides plants with a means of modulating the dissipation of excess energy.

Considerable work on the molecular biology, particularly cloning of the enzymes, VDE and ZE has been done by Yamamoto et al. (1999). Recently, cDNAs for both the enzymes, VDE and ZE, were isolated and the activities of the expressed proteins demonstrated. Based on n-terminal sequence of purified VDE from lettuce which indicates transit peptide cleavage site, mature VDE protein is encoded by 348 amino acids with a molecular mass of 39.9 kDa, close to the reported molecular mass of 43 kDa for purified VDE from lettuce and spinach. The VDE gene is located on chromosome 1 in *Arabidopsis*. Higher levels of VDE mRNA were detected in mature green leaf tissue, suggesting a higher expression in tissues with a higher density of fully-developed chloroplasts (Yamamoto et al. 1999).

To detect the impact of reduced level of VDE, an antisense construct was prepared using tobacco VDE cDNA and integrated in the tobacco genome. A total of 18 plants had various levels of inhibition of de-epoxidation, with two plants having more than 90 per cent reduction in de-epoxidation. cDNA encoding ZE was identified from *Nicotiana plumbaginifolia*. The native ZE protein has not been purified, but evidence exists that ZE as epoxidase of both xanthophyll cycle and the ABA biosynthetic pathway (Yamamoto et al. 1999).

3.3B. Pigments of xanthophyll cycle and their structure

The xanthophyll pigments, VAZ, are uniformly distributed under high light. Violaxanthin is present in both PS I and PS II. The distribution of these pigments in PS II protein complexes has been investigated more thoroughly than that of PS I. V has been found in the pigment proteins except in RC core of PS II (Bassi et al., 1997; Eskling et al., 1997). Eighty per cent of V in maize is said to be located in minor complexes, while bulk LHC II contained the remaining 20 per cent. Therefore, the minor complexes CP24, CP26 and CP29 are considered to posses much higher VAZ/Chl in comparison with bulk LHC II. The fact that VAZ is more enriched in minor complexes CP 29, CP 26 and CP 24 relative to major LHC has been shown by Verhoeven et al., (1999) in their studies of xanthophyll cycle pigment localization in *Vinca major*. The xanthophylls, lutein, violaxanthin and neoxanthin are enriched in LHCs and contribute to the assembly, light harvesting and photoprotection (Horton et al., 1996; Pogson et al., 1998). Mutant studies have shown that the unaltered xanthophyll composition is actually required for optimal assembly and function of light-harvesting antennaa in higher plants.

The amount of V available for conversion is somewhat variable. Maximum V conversion at saturating light is 50 – 80 per cent high degree of conversion is reported in plants deficient in LHC, compared to wild type. A hundred per cent de-epoxilation in pea plants is known under high light with CP 26 as the only PS II antenna pigment protein present. Therefore, the degree of conversion of V to Z is dependent on the amount of pigment-binding proteins. A high degree of conversion was associated with plants deficient in LHC, compared to wild type (Lockstein, et al., 1994; Farber and Jahns, 1998). It is also known that xanthophyll cycle carotenoids are loosely bound to the major complexes. It is suggested by the work of Bugos et al., (1998) that V is to be actually removed from the complex and inserted into the active site of the enzyme for de-epoxidation. Bugos et al., (1999) have demonstrated in experiments conducted on tobacco leaves that the level of VDE enzyme changed inversely with a non-linear relationship in respect of VAZ pool. This observation suggested that the enzyme levels are indirectly regulated by the VAZ pool. Apparently, the bound violaxanthin is not available for de-epoxidation (Ruban et al., 1999). It is suggested that only a small fraction of Z which is required for NPQ and is associated with minor complexes (Gilmore, 1997; Ruban et al., 1999). Ruban et al., (1999) suggested, based on their studies on spinach, that the majority of zeaxanthin pool is associated with LHC II b. Accordingly, the control of NPQ is not restricted to minor complexes, but to the entire LHC II antennae. However, this view was not held by Ruban and Horton (1994). Phillip and Young (1995) observed that 50 per cent of V in PS II was located in the bulk LHC II and only 30 per cent in minor complexes. Accordingly, there appeans to be considerable difference of opinion on the actual distribution of VAZ pigments, an area which seems to be requiring further attention.

CP24, CP26 and CP29, which bind the xanthophyll cycle pigments, have a pericentral location in PS II units and are, thus, seen as connecting antennae between major LHC II and PS II core (Jansson, 1994; Yamamoto and Bassi, 1996). The protein subunits of PS II core comprise, the Chl a/β-carotene complexes and are encoded by the chloroplast genome. The antennae system subunits, which are chl a/b xanthophyll complexes, are nuclear encoded.

The three pigments that participate in xanthophyll cycle are derived from β-carotene and all of them possess one hydroxyl moiety for each cycle end group. Two epoxide groups occur in violaxanthin, one at each cyclic end group. The intermediate xanthophylls — antheraxanthin between violaxanthin and zeaxanthin — possesses one epoxide group only (Gilmore, 1997). The special significance for the de-epoxidized end group structure of zeaxanthin lies in the fact that they serve as binding sites to the protonated CP complexes. Interaction of xanthophylls with a β cyclic end group structure with CP is known to cause a confirmational change affecting the PS II core and inner antennae holocomplex. The number of conjugated double bonds in the xanthophylls has considerable importance in the NPQ mechanism (Demmig–Adams et al., 1996; Frank et al., 1994; Gilmore 1997). It is thought that the increasing conjugation length would render the xanthophyll molecule to accept energy from Chl a and the xanthophyll subsequently dissipates the same energy as heat. The structural configuration is of high significance due to the specific binding site on CPs which accept the de-epoxidized end group and, accordingly, making molecular contact between chlorophyll and xanthophylls.

The 3-D structures of carotenoids have been determined at least for a few molecules with NMR spectroscopy (Mo, 1995). Horton et al., (1999) have discussed the significance of the structure of the carotenoids involved in the xanthophyll cycle in higher plants and algae. The biological role of xanthophylls is obviously dependent on the physico-chemical properties of carotenoid molecules (Britton, 1995; Ruban et al., 1993). It is a well-known fact that the energy content of S1 state of carotenoids decreases with increasing number of conjugated double bonds. DeCoster et al., (1992) have suggested that S1 of highly conjugated carotenoids lies below that of chlorophyll Q_y, as already mentioned above. Conversion of violaxanthin to zeaxanthin results in increased conjugational length to 11 double bonds and removes the epoxides. Thus, S1 state of zeaxanthin has a lowered energy level below that of chlorophyll a (Owens, 1994; Frank et al. 1994).

3.3C. Role of Xanthophyll Cycle

The role of zeaxanthin and that of the xanthophyll cycle in photoprotection is based on chlorophyll a fluorescence quenching measurements. The xanthophyll cycle in vivo is triggered by increased pH gradient across the thylakoid membrane in excessive light. A correlation between EEE and the xanthophyll cycle is also evidenced by the fact that plants produce an increased pool size of xanthophyll cycle pigments when transferred from limiting light to excess light (Demmig - Adams et al. 1996).

The amount of zeaxanthin and the non-photochemical quenching are well correlated. Besides its primary role in NPQ, xanthophyll cycle is considered to participate in several other functions, including protection of lipids against oxidative stress, regulation of membrane fluidity, control of absciscic acid synthesis and possibly in a blue light response.

Non-photochemical quenching (NPQ) refers to the component of fluorescence quenching unrelated to the photochemistry process. That zeaxanthin has a photoprotective function through dissipation of excess energy, is established by the correlation between increased NPQ and zeaxanthin formation under high light conditions (Demmig- Adams et al., 1996; Pfundel and Bilger, 1994 Eskling et al., 2001). This is further evidenced by the correlation of the sum of A and Z to NPQ.

The mechanism of quenching is still not clear. There are two models proposed (Ruban et al., 1998). One is an indirect model, which involves proton-induced structural changes leading to the formation of dissipating centres in specific antennae complexes by antheraxanthin and zeaxanthin. Such structural changes are said to be inhibited by violaxanthin (Ruban et al., 1997, Wentworth et al., 2000).

The direct model involves transfer of energy from violaxanthin to chlorophyll, while chlorophyll transfers energy to zeaxanthin (Frank et al. 1994). Therefore, violaxanthin participates in light harvesting while zeaxanthin is active in energy dissipation. The S_1 state of violaxanthin should be above that of chlorophyll Q_y and S_1 of zeaxanthin should be below Q_y according to this model. Recently, studies conducted on the energy levels of lowest excited singlet states of V and Z have been determined by Frank et al., (2000) through fluorescence spectroscopy. The researchas suggested from their results that a direct quenching of Chl fluorescence by the xanthophylls is energetically feasible. Therefore, the direct model may have a significance in the quenching process. The dissipation of excess energy by xanthophylls is controversial, with two prevailing hypothesis, as already mentioned above. The direct role proposed for the de-epoxidized xanthophyll (Z) involves formation of quenching complexes with light harvesting complexes (Gilmore, 1997). The alternate hypothesis supported by Hortons group has an indirect role. However, there is a general consensus that xanthophyll cycle provides plants with a means to modulate dissipation of excess energy. Further discussions on the mechanism has been further reviewed. Horton et al., 1999; Eskling et al, (1997, 2001).

The actual site of NPQ has also been intensively studied. It was suggested by Gilmore et al., (1996) that the minor CP proteins in PS II as the site of NPQ. This was suggested since LHC II is not required for zeaxanthin quenching. Mutant studies using *Arabidiopsis* by Li et al., (2000) showed that when the chlorophyll-binding protein CP 22 (PsbS) is lacking in the mutant npq 4-1, there was normal xanthophyll cycle and the pigment composition. However, the mutant was incapable of dissipation of excess absorbed light and, therefore, a derangement of NPQ. PsbS was suggested to be the site for quenching (Li et al. 2000). The results of Li et al., (2000) further showed that PsbS protein contributes to energy dissipation and photoprotection than for light harvesting. They suggested a model withn PsbS in the antennaa of PS II as the site of qE and Xanthophyll-dependent quenching. Protonaction of PsbS is believed to lead to quenching of singlet excited chlorophyll. Thiele and Krause (1994) have clearly shown with their studies on isolated thylakoids of spinach (*Spinacia oleracia* L.) that zeaxanthin action in fluorescence quenching and thermal energy dissipation requires acidification of lumen by energization of thylakoid membrane. It was also suggested that in vivo zeaxanthin by stimulating the qE process indirectly protects against photoinhibition.

The photoprotective roles of zeaxanthin and the related aspects of molecular genetics have been discussed relatively recently (Baroli and Niyogi, 2000). Several studies have, on the other hand, showed that the photoprotective function of zeaxanthin is not yet confirmed. Thiele and Krause (1994) have suggested, using spinach, that zeaxanthin serves as a photoprotector only in the energized thylakoid system by stinulating qE process and is not a quencher of excessive energy per se. Some mutants of *Arabidopsis* and *Chlamydomonas*, which are unable to convert zeaxanthin to violaxanthin, have normal photosynthesis for the survival in excess light. These plants do not require the operation of xanthophyll cycle (Hurry et al., 1997; Niyogi et al., 1997). It has also been shown that xanthophyll cycle contributes to the photoprotection by an additional mechanism which is different from energy dissipation (Havaux & Niyogi, 1999; Havaux et al., 2000,). Therefore, much further work is needed in relation to the specifc role of zeaxanthin.

At the entire plant level, zeaxanthin-dependent thermal dissipation increases with increased excess of light (Demmig – Adams et al., 1999). Xanthophyll-cycle dependent energy dissipation has been shown to mitigate the increased levels of reactive oxygen species and triplet chlorophyll under high growth PPFD. The ecophysiology of the xanthophyll cycle has been recently reviewed by Demmig – Adams et al. (1999).

3.3D. Prevention of Lipid Oxidation Stress

Zeaxanthin may act as a quencher of triplet chlorophyll (^{3}Chl) and various reactive oxygen speices (Schindler and Lichtenthaler, 1996). It was shown by Havaux and Niyogi (1999) that in mutants lacking VDE, lipid peroxidation was much higher. Their studies indicate a definite role for xanthophyll cycle in the protection of lipids.

Zeaxanthin has been shown to affect membrane fluidity in the peripheral region of hydrophobic core and cause decreased fluidity. It may act like cholesterol and α-tocopherol in decreasing the membrane fluidity. Havaux (1998) has discussed the role of carotenoids in the stability of membranes. Marin et al., (1996) have exhibited that zeaxanthin is an intermediate in the synthesis of plant hormone, abscisic acid. It has also been shown that under mild water stress, an increase in zeaxanthin has also been recorded, while under severe conditions of water stress, an increased conversion of violaxanthin to zeaxanthin was observed.

Grasses et al., (2001) have studied the performance of transgenic tobacco under high light stress. They have observed that the reduction in qE is not responsible in increasing qI, the photoinhibitory component. The reduction of qE was attributed to a result of increased qI after prolonged illumination. Grasses et al., (2001) have studied the photosynthetic performance and susceptibility to photo-oxidative stress in the transgenic tobacco plants. Their observations state that a reduction in the activity of the enzyme geranylgeranyl reductase in transgenic tobacco plants has produced lowered levels of total chlorophyll and tocopherol but with an increased content of geranylgeranylated Chl (Chl GG).

The data (Grasses et al., 2001) suggest that the presence of Chl GG has no influence on the harvesting and transfer of light energy in either photosystem. But the reduced tocopherol content is found to be a limiting factor in preventing photo-oxidative stress. The increased VAZ pool size can be interpreted as an adaptation to increased light sensitivity. Xanthophyll cycle pigments, in addition to their dissipating function in antennae, also play a role in membrance fluidity and against lipid peroxidation. These functions of xanthophyll resemble those of tocopherol so that a reduction in tocopherol content could be compensated by increase in xanthophyll cycle pigments.

Several phenomena — including phototropism, leaf solar tracking and stomatal opening — are blue light responses and these responses have been shown to be correlated with the content of Z. Srivastava and Zeiger (1995) have observed that applicaton of DTT to the epidermal peels of *Vicia faba* has resulted in suppression of stomatal opening. The role of zeaxanthin as a blue light photoreceptor in the stomatal guard cells has been shown by Frechilla et al., (1999) based on their studies on zeaxanthin — less *Arabidopsis* mutant. However, there are also reports suggesting that zeaxanthin does not function in blue light response (Palmer et al., 1996) for phototropism in maize coleoptiles. Thus, the role of zeaxanthin as a blue light photoreceptor is yet to be established. Yet another role for zeaxanthin (Z) has been demonstrated in the photoprotection of PS II by Jin et al., (2003) using a xanthophyll aberrant mutant, (zeal) of *Dunaliella salina*. These researchers observed that a correlation existed between the xanthophyll cycle and the PS II repair cycle and it was suggested that Z is a component of the PS II repair process. The actual site of protection by Z was considered to be after PS II photodamage and disassembly has occurred during photoinhibition and before recovery and reconstitution of PS II holocomplex (Jin et al. 2003).

It recent years, the existence of a new xanthophyll cycle involving de-epoxidation of lutein epoxide and epoxidation of lutein in certain parasitic plants and also in *Quercus* (Garcia – Plazaola et al., 2003) has been reported. The authors suggested a different function and regulation than the VAZ cycle. However, detailed research is required to assess the relative significance of the VAZ cycle and that of lutein epoxide cycle in relation to energy dissipation.

3.4A. A Reactive Oxygen Species (ROS) : Production of Active Oxygen

As stated above, plants suffer from light stress when light absorption exceeds its demand for photosynthesis. Photoproduction of radicals and active oxygen is unavoidable even under favourable conditions for photosynthesis under their prompt scavenging, is necessary for the protection of target molecules (Asada 1996). It is already established that excess light is harmful to photosynthetic apparatus in view of its action in generating highly reactive oxygen species (ROS) (Asada, 1996, 1998). Hydroxyl radical, super oxide and H_2O_2 can be generated in PS II complexes (Jacob and Heber, 1996; Foyer and Noctor, 2000). An excess of electrons can still result in the photosynthetic electron transport chain, inspite of efficient dissipation of excess light energy in PS II. Photoinhibition of PS II may take place from such over-reduction of electron transport carriers and also leading to the increased production of reactive oxygen species, including (O^-_2) and H_2O_2 (Melis, 1999; Niyogi, 1999; Niyogi, 2000).

It can be further understood that the safest thermal dissipation is a predominant mechanism in eliminating EEE. Prolonged exposure to EEE is known to generate ROS including singlet oxygen, superoxide, hydrogen peroxide and hydroxyl radicals (Karpinski et al., 1999, 2001; Mullineaux et al., 2000). When the capacity of antioxidant systems is insufficient to scavenge them, photo-oxidative damage to chloroplast can take place. Besides the reactive oxygen species (ROS), the intermediates of chlorophyll biosynthesis may also create photo-oxidative stress (Apel, 2001). The Active Oxygen Species (AOS) are produced as a result of partial reduction of oxygen. The term AOS is generic and includes free radicals like superoxide (O^-_2) hydroxyl radical (OH^-) and also hydrogen peroxide (H_2O_2) and singlet oxygen (Noctor and Foyer, 1998; Foyer and Harbinson, 1999; Vranova et al., 2002). The terms ROS and AOS are used here interchangeably. The ROS can cause oxidative damage to proteins, DNA and lipids and the damage is further enhanced by abiotic stresses, including excess light (Smirnoff, 2002). It is also evident that studies in transgenic plants with overexpression of antioxidant enzymes including superoxide dismutase, glutathione reductase and ascorbate peroxidase have been done. It is a well-known fact that a reduction of levels of glutathione in mutants or transgenics reduces the tolerance to stress (Xiang et al., 2001); Meyer and Fricker, 2002). Similarly, an increased production of antioxidant molecules, including glutathione α-tocopherol and ascorbate, was attempted through regulation of their biosynthetic pathways (Grusak and Della Penna, 1999; Smirnoff, 2001, 2002). The elevated levels of glutathione have led to a reduction of stress (Gullner et al., 2001), while some times it may lead to even greater oxidative damage (Creissen et al., 1999). Also, an overexpression of ferretin, an iron storage protein, did not protect from oxidative stress (Murgia et al., 2001). Thus, the role of elevated levels of antioxidants in the protection of oxidative stress is controversial. An antioxidant denotes any compound which is capable of quenching AOS while itself not undergoing conversion to a destructive radical (Nishikimi and Yagi, 1996). Recently, 2-cysteine peroxiredoxins (2-CP) were also identified as members of the antioxidant defense system of chloroplasts (Dietz et al., 2002). Reactive oxygen species and peroxides are cell toxic and also indicators of the metabolic performance of cells. They may act as components of signalling pathways. During photosynthesis, AOS are produced in a concerted manner and contribute to the regulation of electron transport (Asada, 1992, 1997).

3.4B. Antioxidant System in Scavenging Active Oxygen

Plants have developed a range of protective mechanisms to mitigate the deleterious effects of excess light by either scavenging toxic activated oxygen species or preventing their production (Anderson et al., 1997). An increase in the expression of antioxidant genes including ascorbate peroxidase and dehydroascorbate reductase (DHAR) and several unknown genes — either regulatory or of metabolic enzymes — was recently demonstrated by Rossel et al., (2002) in *Arabidopsis*. This increased gene expression was caused when plants of *Arabidopsis* were transferred from a light intensity of 100 µ mol $m^{-2}s^{-1}$ to a high light of 1000 µ mol $m^{-2}s^{-1}$ for 1hour. Therefore, plants respond effectively to oxidative stress and display increased reased level of antioxidants to mitigate stress. Both enzymatic and non-enzymatic components exist in higher plants related to the detoxification of ROS formed already. Such components include antioxidant non-enzymatic small molecules, ascorbate, glutathione and α-tocopherol (Fryer, 1992). It has been shown that reduced ascorbate and reduced glutathione are involved in preventing photo-oxidation and providing photoprotection (Augusti et al., 2001). However, the results drawn by Xu et al., (2000) have shown in rice leaves that exogenous application of glutathione has suppressed the formation of A and Z and also inhibited NPQ. The results suggested that the action of GSH is on xanthophyll cycle. The lipid soluble antioxidant α-tocopherol is located in the thylakoid membranes and counteracts the effects of ROS (Gonzalez – Rodriguez et al. 2001). The enzymes, including superoxide dismutase, catalase, glutathione reductase and ascorbate peroxidase, are the other group of antioxidants, (Logan et al., 1998; Vranova et al., 2002). Ascorbate peroxidase exists in isoenzymes and plays an important role in the metabolism of H_2O_2 in higher plants (Shigeoka et al. 2002).

The components of the antioxidant system can be classified in the following manner:

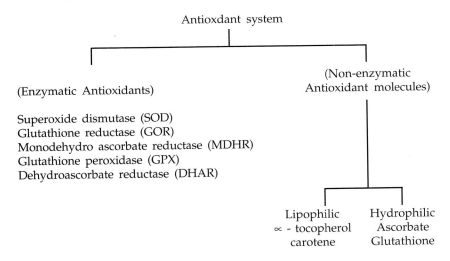

Besides the classical antioxidant system, a phenylpropanoid pathway is proposed by Grace and Logan (2000) for radical scavenging and energy dissipation. The phenolic function in offering tolerance to light stress with particular reference to chlorogenic acid has been reviewed. It was also suggested (Grace and Logan, 2000) that the phenylpropanoid pathway may provide an alternate route for photon use under excess light. This study requires further elaboration in relation to the existing pathways of antioxidant production and their roles in radical seavenging.

Recent results indicate that enhanced susceptibility to AOS is not necessarily a result of reduced NPQ but is due to a loss in capacity for destruction of singlet $^1O_2$. VDE deficient mutant, i.e is NPQ 1-1 can exhibit enhanced peroxidation in strong light. Conversely, the NPQ deficient PS II-S deletion mutant retains the normal xanthophyll cycle and exhibits high tolerance to excessive light. Therefore, evidence does not support the theory that singlet oxygen, in the transformation of PS II centres from active to inactive form in NPQ 4-9 forms a new mutant of *Arabidopsis thaliana* with reduced NPQ. At this time, it is unclear as to how NPQ 4-9 alters both NPQ mechanism and the control of electron flow between PS II and PS I (Peterson and Havir, 2001).

Recently, Kulheim et al., (2002) have investigated the role of qE or the ΔpH dependent NPQ feedback de-excitation in the fitness of *Arabidopsis* mutants in different light environments. They have shown that feedback de-excitation was crucial for plant performance in field conditions or in fluctuating light conditions. It apparently does not affect the fitness under steady state conditions. The authors (Kulheim et al., 2002) concluded that feedback excitation confers particular adaptive advantage as a short-term regualtion of photosynthesis but not a protective mechanism against high light. It was also shown that zeaxanthin-dependent lipid protection is not so important as the feedback de-excitation. This work is of considerable significance in view of the fact that studies on NPQ have been extended to the entire plant under field conditions and, therefore, further work in this direction is expected to give better insights into the role of short-term photoprotective measures for plant growth and performance.

The photoproduced reactive molecules cause the damage to target molecules D1 protein of PS II RC and majority of calvin-cycle enzymes which results in photoinhibition of photosynthesis. D1 protein in PS II RC and the Calvin cycle enzymes are the primary target molecules for photoinhibition, which can be suppressed by different mechanisms: photoproduction of reactive molecules; photorespiration; cyclic electron flow around PS I; xanthophyll cycle and state transitions are some of the mechanisms. While the excess energy dissipation takes place through heat, a proportion of photon energy is transferred inevitably to chlorophyll and oxygen, resulting in triplet chlorophyll and singlet oxygen (Asada, 1994).

Reactive molecules include reduced dioxygen including superoxide, H_2O_2 and OH, and excited singlet oxygen. Triplet chlorophyll and MDA radicals are also photogenerated. In order to prevent interaction of ROS with target molecules, their diffusion from the site of generation has to be controlled. Therefore, the plants posses systems to scavenge the ROS before they diffuse from the site of generation (Asada, 1994). Superoxide radicals are disproportionated at appreciable rates.

There are two sites of scavenging systems: the thylakoid and the stromal systems. Ascorbate peroxidase catalyses reaction, reducing hydrogen peroxide in water and produces MDA.

The sequence of reactions involved in the elimination of the reactive oxygen species is as follows :

a. Reduction of O_2 at PS I to O^-_2
b. Dismutation of O^-_2 to H_2O_2
c. Reduction of H_2O_2 to H_2O
d. Regeneration of ascorbate

This sequence constitutes the Mehler – Peroxidase cycle (Polle, 1996; Foyer et al., 1997). Recently, Polle, (2001) has studied the superoxide dismutase – ascorbate – glutathione pathway in chloroplasts by metabolic modelling. The removal of H_2O_2 through the series of reactions is known as ascorbate – glutathione cycle (Karpinski et al., 1997; Polle, 1997; Noctor and Foyer, 1998; Asada 1999). A schematic representation of Mehler - Peroxidase reaction and that of Ascorbate – glutathione cyclical process is shown in Figures 3.3 and 3.4. Ascorbate peroxidase uses two molecules of ascorbate to reduce H_2O_2 to H_2O with the production of 2 moles of MDHA (MDA) (radical with short life span) MDA disproportionates to DHA. DHA is reduced to ascorbate through DHA reductase using GSH as substrate producing glutathione disulfide (GSSG). Then GSSG is reduced by NADPH through glutathione reductase. That oxygen reduction may also take place in a phastoquinone pool in the isolated thylakoid membrane has been shown by Khorobrykh and Ivanov (2002). The significance of these findings in relation to Mehler reaction in vivo needs to be further examined.

The antioxidative systems are crucial for the detoxification of AOS. Ascorbate is the most soluble antioxidant in the chloroplast. A superior antioxidant capacity has been demonstrated in *Quercus ilex* as an adaptive strategy to overcome high light and low temperature stress (Garcia – Plazaola et al. 1999). Smirnoff (2000) has made an important study in identifying the multiple roles of ascorbate in photosynthesis and photoprotection. Biosynthesis of ascorbate is shown to be controlled by light. The biosynthetic pathway for ascorbate in plants is a characteristic process, which it has been elucidated recently (Wheeler et al. 1998; Smirnoff, 2000). α-Galactose the precursor it is oxidized to L - galactano– 1, 4-lactone by an NAD dependent L - galactose dehydrogenase located in the inner mitochondrial membrane. Relatively high levels of

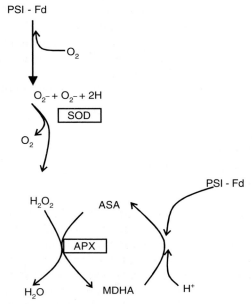

Fig. 3.3. The Mehler–peroxidase system. Reduced ferredoxin when reacts directly with oxygen, hydrogen peroxide is generated via superoxide formation. H_2O_2 is eliminatd in presence of ascorbate peroxidase. SOD = Super Oxide Dismutase; ASA = Ascorbate, MDHA = Monodehydro ascorbate; Fd = Ferredoxin.

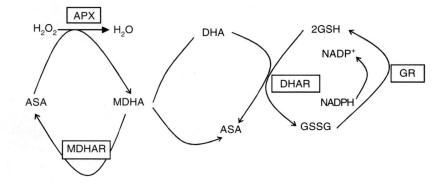

Fig. 3.4. Ascorbate –glutathione cycle. Monodehydroascrobate produced in the ascorbate peroxidase reaction is reduced to ascorbate by monodehydroascorbate reductase. Dehydro ascorbate generated due to disproportionation of monodehydroascorbate is reduced back to ascorbate through the involvement of glutathione. APX = Ascorbate peroxidase; MDHA = Monodehyroascorbate; MDHAR = Mono dehydroascorbate reductase; DHAR = Dehydro ascorbate reductase; GSH = Glutathione reduced; GSGG = Glutathone disulphide; GR = Glutathione reductase.

ascorbate are essential under conditions of extreme levels of photo-oxidative stress. It was shown earlier (Logan et al., 1998) that higher levels of antioxidants responsible for scavenging reactive reduced oxygen species were found in plants growing in high light environments as compared to those in lower light regimes. The high light plants also exhibited high levels of xanthophyll cycle dependent energy dissipation (Logan et al.1998).

Glutathione

This antioxidant is a tripeptide and exists interchangeably with GSSG (oxidized) (Noctor & Foyer, 1998). The biosynthesis of glutathione involves two ATP-dependent steps catalyzed by γ-glutamyl cysteine synthetase (γ-ECS) and glutathione synthetase (GS). Studies have shown that the content of foliar glutathoine is high in light. Glutathione synthesis is one pathway that could utilize the intermediates of photorespiratory cycle, since glycine is formed in significant quantities during photorespiration.

The scavenging of active oxygens is facilitated by water–water cycle which is a sequence of reactions involving the photoreduction of dioxygen to water through formation of super oxide and hydrogen peroxide at the site of PS I by electrons originating from water in PS II (Asada 1999). The chloroplast O_2 – photoreducing system to water is termed as water – water cycle, since electrons from water in PS II are diverted to oxygen for reduction to water at PS I without net oxygen exchange. The pathway is of considerable importance in the protection from photoinhibition. Mehler (1951) found the photoreduction of O_2 to H_2O_2. The photoreduction site of oxygen was shown to be PS I and its product was identified as super oxide anion radical (Asada et al. 1974).

Although the term Mehler – Peroxidase reaction was used for this cycle, due to non-inclusion of steps in regenerating ascorbate, the term water – water cycle is preferred (Asada, 1999). The crucial function of this cycle is rapid and immediate, scavenging O_2^- and H_2O_2 at the site of production before they interact with target molecules. The cycle has a dual function of scavenging active oxygens and safe dissipation of excess photon energy. It is said that 30 per cent of the linear electron flux passes through the water – water cycle. Therefore, this cycle provides a safety valve (Niyogi, 2000) to dissipate EEE under environmental stress. Under excess photons, the physiological electron acceptors are not available to PS I. The proton gradient is generated either by cyclic electron flow around PS I or through the water – water cycle. The water – water cycle dissipates excess photons using oxygen as an electron acceptor.

The two physiological functions have been ascribed to water – water cycle are :

(a) Protection of scavenging enzymes, the stromal enzymes and the

PS I complex from oxidative damage by O_2^- and other ROS produced in PS I.
(b) The reinforcement of CO_2 assimilation by supplying ATP for the carbon reduction cycle.
(c) Also, a third function comprises dissipation of excess photons.

Miyake and Yokota (2000) have studied the electron fluxes for the PCR and PCO cycles and the photoreduction of O_2 at PS I. It was found that in water melon (*Citrulus lanatus*) leaves, the total electron flux in PS II was larger than that required for PCR and PCO cycles. Therefore, the existence of an alternative electron flux was suggested. The alternate flux was ascribed to the water – water cycle, though there is no direct proof that this cycle operates at high rates. The study by Miyake and Yokata has indicated that there is a significant water – water cycle activity (Miyake and Yokota, 2000, Ort and Baker, 2002). Superoxide radicals produced at PS I are disproportionate to H_2O_2 and O_2 by SOD. H_2O_2 which accumulates in stroma and inactivates PCR cycle enzymes.

The water – water cycle (Fig. 3.5) operates through the following sequence of reactions.

$$2 H_2O \longrightarrow O_2 + 4H + 4e^- \text{ (PS II)}$$
$$2e^- + 2 O_2 \longrightarrow 2 O_2^- \text{ (PS I)}$$
$$2 O_2^- + 2 H^+ \longrightarrow H_2O_2 + O_2 \text{ (SOD)}$$
$$H_2O_2 + 2 \text{ AsA} \longrightarrow 2 H_2O + 2 \text{ MDHA (APX)}$$
$$2 \text{ MDHA} + 2 e^- H^+ \longrightarrow 2 \text{ AsA (Fd or MDHAR)}$$

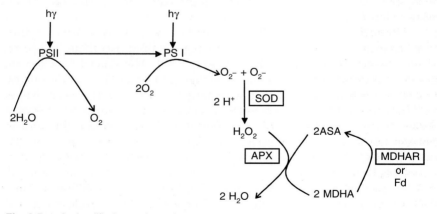

Fig. 3.5. A simplified overview of water – water cycle. Electrons from water molecule enter PS II. Oxygen is photoreduced at PS I through Mehler Reaction. H_2O_2 is reduced to water through the enzyme system of ascorbate peroxidase and monodehydroascorbate reductase. SOD = super oxide dismutase; APX = Ascorbate peroxidase; ASA = Ascorbate; MDHA = Monodehydroascorbate; MDHAR = Monodehydro ascorbate reductase.

O_2^- which is produced at PS I is disproportionate to O_2 and H_2O_2. This reaction is catalysed by superoxide dismutase (SOD) and, subsequently, ascorbate peroxidase (APX) reduces H_2O_2 to water using ascorbate (AsA) as an electron donor. The monodehydroascorbate radical (MDHA) produced is reduced back to AsA by either ferredoxin (Fd) or by NADH catalysed by monodehydroascorbate reductase (MDHAR).

While there is disadvantage by photoreduction of O_2 at PS I, it is important for protection of PS II from photoinhibition.

3.5. ADDITIONAL ELECTRON SINKS

It is known that the maximal level of NPQ reaches before light saturation and therefore, only photochemical quenching should be responsible for photoprotection at saturating light intensities (Ort and Baker, 2002). The photochemical quenching of EEE refers to the consumption of electrons by additional metabolic sinks (Mullineaux and Karpinskii, 2002). These additional sinks include the reduction of oxygen at PS II or PS I, enhanced rates of photorespiratory and chlororespiratory metabolism and to a limited extent increased C, N and S metabolism (Fryer et al., 1998; Streb and Feierabend, 1999; Mullineaux and Karpinskii, 2002). The electron transfer to oxygen prevents over-reduction of the electron transport chain and thus reduces the risk of harmful reactions. It is presently understood that oxygen has a substantial role by serving as an alternate electron acceptor in photoprotection under high light intensities and environmental stresses (Ort & Baker, 2002). Oxygen is crucial as an alternative electron sink. Oxygen reduction is regarded to contribute substantially for the photosynthetic electron flux through participation not only in photorespiration but also by direct reduction at PS I (Asada, 1996). Palatnik et al., (1999) while studying the oxidative degradation of chloroplastic glutamic synthetase in wheat, have demonstrated that oxygen may serve as a sink of surplus electrons under limitations of CO_2 assimilation. Asada (1999) has suggested that photoreduction of oxygen to water by Mehler – Ascorbate peroxidase pathway may constitute as much as 30 per cent of the total electron flux. It has been repeatedly proposed earlier, that electron transfer to oxygen is an effective means for prevention of excessive reduction of electron transport chain (Biehler and Fock, 1996; Osmond and Grace, 1995; Ott et al., 1999; Heber et al., 2001). The Mehler reaction itself is viewed from different angles (Ott et al., 1999). One of the views is that superoxide formation is harmful and damaging and requires heavy investment of antioxidants, including ascorbate and, therefore, minimizing superoxide formation is beneficial (Genty and Harbinson, 1996). On the other hand (Polle, 1996), it is believed that oxygen serves as an alternative source for electrons and reduce the pressure on electron transport chain originating from PS II. It also produces Δ pH gradient for other protective processes (Polle, 1996). However, contrary to this view, oxygen reduction during Meher reaction is considered to be rather slow (Heber et al., 1995; Wiese et al., 1998; Ruska et al., 2000; Clarke and Johnson, 2001). It is understood that Mehler reaction (linear electron flow to oxygen) alone is incapable of decreasing intrathylakoid pH appreciably for the control of PS II, while cyclic electron transport is effective in this regard. Thus, the role of oxygen appears to be a redox poising than as an electron acceptor. A crucial physiological role is assigned to photorespiratory metabolism in this connection (Krause, 1988; Heber et al., 1995; Kozaki and Takeba, 1996; Wiese et al., 1998; Heber et al., 2002).

Cyclic electron transport has earlier been assigned a role in the photoprotection of leaves against photoinhibition (Heber and Walker, 1992; Gerst et al., 1995). Cyclic electron transport around PS I is regarded to play a role in the generation of ATP (see Figure 2.1). In the cyclic electron flow around PS I, electrons are returned from PS I to the linear chain at the site of plastoquinone pool or cytochrome b6f complex (Howitt et al., 2001). In general the cyclic electron transport has two main pathways (Joet et al., 2001; Barth and Krause, 2002). One of these pathways is sensitive to antimycin A (Ivanov et al., 1998) and uses ferredoxin plastoquinone oxidoreductase, while the other, insensitive to antimycin A, involves the thylakoid NAD (P) H reductase (NDH) complex (Joet et al. 2002). It was demonstrated by Joet et al., (2002) that a high and rapid cyclic electron flow around PS I exists in vivo in C_3 plants. Also, it was clear that rapid cyclic transport around PS I related to NDH complex occurs under anaerobiosis. Cyclic electron flow around PS I produces Δ pH while the linear electron transport through both the photosystems results in NADPH and Δ pH. Accordingly, when the balance of linear and cyclic electron flows is changed, the H^+/e^+ ratio could be altered. A regulation results in increased Δ pH and promotes thermal dissipation (Cornic et al., 2000, Shikanai et al., 2002). Cornic et al., (2000) have investigated the role of cyclic electron transport in leaves of C_3 plants. Their studies indicate that cyclic electron flow regulates the quantum yield of PS II by lowering the intrathylakoid pH under the conditions of environmental stresses with lowered levels of electron acceptors. Further, it was suggested the operation of Q-cycle is not obligatory but only flexible. In respect of requirement for enhanced photoprotection during winter stress in evergreens, there is a retention of PS I photochemistry and also there was enhanced cyclic electron transfer around PS I in the cold-stressed needles of Scots Pine (Ivanov et al. 2001). Thus, the energy supplied through cyclic transport is utilized for the recovery of photosynthesis from cold stress.

Cyclic electron flow within PS II (CEF – PS II) has also been regarded as a mechanism for alleviation of photoinhibition (Miyake et al. 2003). It was shown that the activity of CEF – PS II increases with a corresponding increase in light intensity and may exceed that of the water-water cycle (Miyaka et al., 2002), while the activity of CEF – PS II depends on Δ pH, it is driven by a different mechanism other than that of the dependence of NPQ on Δ pH. Miyake et al., (2003) have concluded that if electron flow in PS II is kept high by CEF – PS II or linear electron flow, photoinhibition of PS II is suppressed (See Figure 2.1).

Photorespiration could be regarded as a substantial alternative sink for electrons (Osmond and Grace, 1995; Streb et al., 1998; Wingler et al., 2000). Oxygenation of RUBP consumes oxygen. Chlororespiration is perhaps another-oxygen dependent pathway in chloroplasts serving as a alternative sink (Niyogi, 2000; Ort and Baker, 2002). A respiratory chain transferring electrons from NAD (P) H to oxygen *via* plastoquinone pool was proposed by Bennoun (1982). Considerable experimental evidence, suggesting that chlororespiratory non-photochemical plastoquione reduction and plastoquinol oxidation occurs in the chloroplasts of higher plants, was provided by Field et al., (1998) using sunflower leaves. Genes encoding (Ndh) complex are located in chloroplast genome. The enzyme activity was found in the chloroplasts (Sazanov et al. 1998). Casano et al., (1999) have determined the levels of antioxidant protective enzymes on the photo-oxidative stress and have observed a close relationship between the actions of Ndh complex and peroxidase. Later, Casano et al., (2001) have studied the induction of chloroplastic Ndh complex in barley in relation to photoxidative stress. They have proposed that the induction of chloroplastic Ndh genes is mediated by H_2O_2 through a rapid translocation of pre-existing transcripts and increased Ndh transcript levels. Cyclic electron transport was defective in tobacco mutants lacking Ndh complex. Endo et al., (1999) have investigated the role of chloroplastic NAD (P) H dehydrogenase and observed that the Ndh defective mutants had higher sensitivity to photoinhibition. Ndh complex has a dual function (Burrous et al. 1998). It functions to catalyse cyclic electron flow around PsI, while in the darkness, it functions in chlororespiration (Nixon, 2000; Nixon and Mullineaux, 2001). Barth and Krause (2002), using tobacco transformants, have shown that the cyclic electron flow operating through the Ndh complex does not presumably contribute to photoprotection of either PS I or PS II against short-term light stress, while this type of cyclic electron transport is perhaps more related to long-term acclimation stress.

The existence of chlororespiratory oxidase has been established in *Arabidopsis* from a study of *immutans*, a variegated mutant (Carol et al., 1999; Nixon, 2000). The IM gene encodes a protein similar to alternative oxidase of mitochondria. The IM protein is involved in phytoene desaturation (Carol et al., 1999) since mutations in this gene accumulate phytoene and block carotene biosynthesis and the protection against photo-oxidation is suppressed. The chloroplastic alternative oxidase is responsible for the removal of electrons from plastoquinone pool. Its activity is critical for carotenoid biosynthesis. It is regarded as providing a safety valve for excess electrons in high light (Niyogi, 2000).

The process of chlororespiration has been reviewed recently by Peltier and Cournac (2002) who defined it as respiratory electron transport chain interacting with photosynthetic ETC in thylakoid membranes. The plastid-encoded (Ndh) complex and the nuclear encoded terminal oxidase (PTOX) are involved in the process. They have presented work related to understanding the role of chlororespiration, suggesting a regulation of photosynthesis through modulation of activity of cyclic electron flow around PS I. PTOX may serve as a safety valve to avoid over-reduction of PS II in high light. Cyclic flow around PS I assists in energy dissipation for protection and repair mechanisms (Finazzi et al. 2001). Transthylakoid pH gradient generated by chlororespiration may help in the process of recovery from photoinhibition (Fischer et al. 1997). It is at present unclear if the contribution of chlororespiration is significant for photoprotection. It is considered to comprise only 0.3per cent of electron flux of light-saturated photosynthesis (Field et al., 1998) and, therefore, further work is needed to determine its role in photoprotection.

Heber et al., (2002) have examined in detail the control of PS II in leaves of higher plants as well as in mosses and lichens. It was suggested that oxygen reduction during Mehler reaction is slow (Badger et al., 2000; Clarke and Johnson, 2001) and further along with the water – water cycle it only poises the electron transport chain for coupled cyclic electron transport than as an electron sink. The transmembrane proton transport generated by photoassimilatory and photorespiratory electron flow is considered to be insufficient for intrathylakoid acidification. Cyclic electron transport is considered to be responsible for production of intrathylakoid acidification, leading to effective dissipation of EEE as heat. In order to assess the roles of H^+/e^- stoichiometry and cyclic electron transfer pathways, Sacksteder et al., (2000) have developed a dark interval relaxation kinetic analysis for quantification of steady–state fluxes of photosynthetic electron system. Since a ratio of H^+/e^- ratio of 3 at low light is usually accepted, it is supposed that this status is maintained under unstressed photosynthesis which suggests a continuous engagement of Q cycle pumping protons at cytochrome b6f complex (Sacksteder et al., 2000). Clarke and Johnson (2001) have studied the rates of electron transport to oxygen in intact barley leaves and examined in detail the potential of Mehler reaction and cyclic electron transport in the control of PS II activity by generating enhanced thylakoid proton gradient. They have concluded that coupled cyclic electron transport is more important in the Δ pH dependent control than electron transport to oxygen in Mehler – reaction. It was further suggested from their results that the slower Mehler reaction, together with photorespiration, is perhaps required to balance the electron transport chain and permit cyclic electron transport to take place to avoid over-reduction.

Among the photoprotective measures, the process of state transition is considered to be important in balancing the light-harvesting ability of the two photosystems (Snyders and Kohorn, 2001). When the absorption of light by Photosystem II is favoured over that from PS I, there is transfer of energy by reversible phosphorylation by LHC P. The mechanism of photosynthetic state transitions has been reviewed by Haldrup et al. (2001). It is a well- known fact that plants can balance the distribution of absorbed light energy between Photosystem I and Photosystem II. A mobile pool of LHC II moves from PS II to PS I when PS II is favoured. The regulation of reversible phosphorylation of LHC II is complex. A series of transgenic *Arabidopsis* plants were created which lack in individual subunits of PS I. Plants lacking the PS I-H or PS I-L subunits were found deficient in state transitions. It is proposed (Haldrup et al., 2001) that other factors besides LHC II phosphorylation might regulate state transitions. This is particularly so in view of the fact that a functional attachment site on PS I is necessary for the attachment of LHC II to PS I and for detachment of LHC II from PS II. Therefore, PS I is considered to be actively involved rather than possessing a passive role in state transitions. Further work is needed in the precise identification of the factors involved.

The overall photoprotection and acclimation to changing environmental circumstances is brought about by several mechanisms which may operate either alone or by multiple pathways alone or at times simultaneous operation of more than one pathway for mitigation of the stress. Much work is needed deeper in understanding of the molecular mechanisms and signal transduction pathways (Mullineaux and Karpinski, 2002) leading to the entire plant acclimation.

4
Leaf Heliotropism, Solar Tracking and Regulation of Light Interception

4.1. INTRODUCTION

Light quantity in natural environments is highly variable, both temporally and spatially (Long et al., 1994; Alves et al., 2002; Oguchi et al., 2003). Such variation in light environment imposes a great demand on the response of photosynthetic system. The plants, therefore, to survive in the highly fluctuating light environments should posses an ability to regulate the level of acquisition of excitation energy both in the long as well as short term.

Changes in light interception can be achieved in several ways. The regulation of light interception through changes in leaf orientation are perhaps the most powerful means of controlling the degree of light interception (Bjorkman and Demmig–Adams, 1995). Leaf movements elicited by the direction of solar beam (Heliotropism) are adaptive mechanisms for two principal functions : (a) photoprotective strategy of light avoidance, thus, mitigating photoinhibition; and (b) maximization of light interception and increased photosynthetic efficiency. Plants possess the ability to respond to various environmental stresses through developmentally-controlled mechanisms. In contrast to such relatively slower responses, the leaf movements are fast and reversible (Yu and Berg, 1994). Changes in leaf angle in response to light environment have long been recognized as a phenomemon common to many land plants. Active leaf movements are observed in species belonging to different taxononic groups which have the ability to actively change their orientation during the day, synchronizing with the solar beam. Such leaf movements are regarded as adaptations leading to the regulation of photosynthetic efficiency (Koller, 1990). Several plants have the ability to adjust their capacity to harvest sunlight through leaf movements (Niyogi 2001). Heliotropism refers to the adjustment of leaf angle and leaf azimuth to the changes in the position of solar beam (Hirata et al., 1983; Meyer and

Walker, 1981; Kao and Tsai, 1998). It is also regarded that heliotropism is a special manifestation of phototropism. Phototropism is related to limiting PAR, while heliotropism occurs under full sunlight and is under the control of the solar beam (Koller, 2000).

Foliar heliotropism, the daily movement of leaves to follow the sun, needs to be differentiated from phototropism, which is turning towards a fixed light source and not tracking the solar beam. Leaf heliotropism occurs in at least 16 plant families (Ehleringer and Forseth, 1980), of which detailed physiological mechanisms of leaf heliotropism have been described only in two families, Fabaceae and Malvaceae (Sherry and Galen, 1998).

The heliotropic responses have been considered to have a dual function of achieving photosynthetic efficiency in utilizing the limited resources and also to reduce the damage due to photoinhibition under excess light (Kao and Tsai, 1998).

The leaf movements in relation to the solar beam can be classified into two categories (Rajendrudu and Das, 1981). The paraheliotropic movements, characterized by the plane of the leaf lamina parallel to incident light rays, decrease the amount of the light interception by the leaf. Therefore, paraheliotropism constitutes an important photoprotective strategy besides its role in the amelioration of low water availability. Diaheliotropic leaf movements, on the other hand orient the plane of the leaf lamina perpendicular to incident light. These leaves obviously maximize the solar interception, which is an opposite strategy to paraheliotropism that is responsible for a light avoidance protective mechanism under high light stress (Donahue and Berg, 1990; Smith 1984; Ehleringer and Forseth, 1980; Sailaja and Das, 1996a). The phenomenon of leaf heliotropism is known for a long period but the mechanisms of perception and the foliar orientation are not clearly understood (Smith 1984). As mentioned above, the maximization of interception of light is achieved by diaheliotropic plants whose leaves possess the ability of tracking the sun by positioning of the laminae perpenducular to the direction of solar beam.

The leaf movements in higher plants in response to the direction of solar beam have been recognized for quite some time but they were viewed only as a mechanism for avoidance of absorption of excess light. A number of species, however, exhibit solar tracking to maximize light interception by the leaf surface in the absence of water stress (Powles, 1984; Ehleringer and Forseth, 1980; Koller and Ritter, 1994; Sailaja and Das, 1996). In contrast to the above situation, a majority of plants are with a fixed leaf orientation in specific compass directions and are known as static leaved plants.

4.2. PARAHELIOTROPISM AND SIGNIFICANCE

As already mentioned above, the phenomenon of paraheliotropism which occurs in a large number of plants including several legumes, results in achieving considerable physiological and ecological advantages. These advantages include an increase in the water use efficiency, a decrease in photoinhibition, increased nitrogen use efficiency and maximization of the photosynthetic rate (Forseth and Ehleringer, 1983; Forseth, 1990; Kao and Forseth, 1992; Cronlund and Forseth, 1995).

External factors preventing the physical movement of leaves alter the absorption of light energy. Paraheliotropic movement results in the maintainance of the leaf parallel to the direction of the sun, minimizing the absorption of solar radiation. By artificially, restraining leaf movements, it was shown that paraheliotropic movement avoids photoinhibition in soybean (*Glycine max*), specifically associated with water stress (Kao and Forseth, 1992; Long et al., 1994).

The paraheliotropic responses are shown to be habitat dependent. The intensity of paraheliotropic movement is variable even in related taxa growing in different environments, because of its advantages related to the specific environments. In this context, Yu and Berg (1994) have conducted a detailed study of paraheliotropism in the speices of *Phaseolus*. They investigated the phenomenon in *Phaseolus vulgaris* from mesic habitat and *P. acutifolius*, an arid inhabiting species. It was observed that *P. acutifolius* exhibited greater paraheliotropism than *P. vulgaris* under the same conditions. It was also clear that paraheliotropism in *P. vulgaris* was observed only under high light levels and high temperatures. *P. acutifolius* of arid habitats achieves high water use efficiency through greater paraheliotropism. *P. vulgaris*, in contrast to the arid species, has reduced paraheliotropic response promoting higher light interception, which is suitable for its growth under mesic habitats (Yu and Berg, 1994). Therefore, there is considerable plasticity in paraheliotropic behaviour related to the characterstic physiological requirements conditioned by the specific growth environment, which does not remain constant. Kao and Tsai (1998) have investigated the pattern of heliotropic movements in three species of wild soybean, *Glycine soja*, *G. tomentella* and *G. tabacina*. Earlier, Kao et al., (1994) have proved that variation exists in leaf orientation in common bean cultivars, which were shown to be related to the their responses to water stress. The studies on soyabean by Kao Tsai, (1998) have supported the earlier view that the paraheliotropic leaf movements result in increased water use efficiency and a reduction in photoinhibition under water stress. They have quantified the movements by the measurement of Cos i to determine the midday leaflet orientation. It was found that *G. soja*, the most suseptible species for drought stress, has the lowest Cos i values, while *G. tabacina* with highest Cos i was the least susceptible species. The *G. tomentella* was found to be an intermediate. Therefore, these studies would provide an evidence for the advantages of para heliotropic movements in achieving higher efficiency under limiting environments.

4.3. DIAHELIOTROPISM

In contrast to paraheliotropic leaf movements, the phototropic leaf movements involving reorientation of the leaf lamina with its upper surface normal to the direction of solar bean are termed diaheliotropic movements (Koller, 1990). The diaheliotropic response is known to maximize the interception of photosynthetically-active radiation (PAR) throughout the day and improves daily total photosynthesis, resulting in higher water use efficiency (Ehleringer and Forseth, 1980; Rajendrudu and Das, 1981; Prasad and Das, 1984; Oosterhuis et al., 1985). The high efficiency of PS II and an apparent lack of photoinhibition in the diaheliotropic plants has been established (Sailaja and Das, 1996a; Sailaja et al., 1997). Several plants, which exhibit leaf diaheliotropic behaviour, are listed in Table 4.2. The uniqueness of photosynthetic behaviour of these plants is discussed later in this chapter. Diaheliotropism is generally viewed as light-seeking behaviour while paraheliotropism is a light avoidance mechanism.

4.4. LEAF HELIOTROPISM AND PHOTOSYSTEM II EFFICIENCY

While the leaf movements in relation to solar beam have been predominantly viewed in terms of avoidance of photodamage through physical means of reduction in light absorption, the PS II photochemistry in these plants has been given a great deal of importance in recent years. The solar tracking plants, known to exist in about twenty angiospermous families, are taxonomically amorphous. The families that contain plants which exhibit one or the other type of heliotropism include: Nyctaginaceae, Amaranthaceae, Euphorbiaceae, Asteraceae, Zyogophyllaceae, Portulacaceae, Martyniaceae, Solanaceae, Fabaceae, Malvaceae and Aizoaceae, among others. A compilation from the literature (Ehleringer and Forseth, 1980; Rajendrudu and Das, 1981; Prasad and Das, 1984; Koller, 1990; Sailaja and Das, 1996a; Koller, 2000; Rao and Das, unpublished) is made and a list of some of the plants exhibiting heliotropic behaviour is shown in Table 4.1.

The phenomenon of leaf heliotropism is quite widespread and has perhaps originated several times in course of evolution. The movement of leaf blade parallel to sunrays is termed paraheliotropic. The

Table 4.1. Some of the plants exhibiting heliotropic behaviour. Compiled from Ehleringer and Forseth (1980); Rajendrudu and Das (1981); Prasad and Das (1984). Koller (1990); Sailaja and Das (1996); Koller (2000); Rao and Das (unpublished) Dia = Diaheliotropism Paro = Paraheliotropism.

Sl. No.	Family	Plant Species	Type of Heliotropism
1	Aizoaceae	*Trianthema portulacastrum*, L.	Dia
2	Amaranthaceae	*Amaranthus viridis*, L.	Dia
3	Apiaceae	*Centella asiatica*, Urb	Dia
4	Asteraceae	*Dicoria canescens*, Tarr and A Gray.	Dia
		Palafoxia linearis, Lag	Dia
		Hetianthus annuus,	Para
5	Boraginaceae	*Coldenia, nuttalii*, L.	Dia
		Heliotropium zeylanicum, Lam	Dia
6	Caesalpinaceae	*Cassia occidentalis*, L.	Para
		Cassia angustifolia, Vahl	Dia
7	Capparidaceae	*Cleome gynandra*, Dc	Dia
		Cleome pilosa	Dia
		Capparis spinosa	Dia
8	Convolvulaceae	*Ipomea eriocarpa*, R.Br.	Dia
9	Euphorbiaceae	*Euphorbia hirta*, L.	Dia
		Euphorbia pulcherrima, Willd	Dia
		Ricinus communis, L.	Dia
		Croton sparciflorus, Mor.	Dia
		Phyllanthus niruri, L.	Para
10	Fabaceae	*Lupinus arijonicus*, L.	Dia
		Astragalus lentiginosus, L.	Dia
		Lotus saluginosus, L.	Dia
		Cicer arietinum, L.	Dia
		Canavalia ensiformis, Dc	Para
		Cajanus cajan, Miisp	Para
		Vigna unguiculata, (L.) Walp	Para
		Vigna radiata,	Para
		Vigna mungo, (L.) Hepper	Para
		Dolichos biflorus, L.	Para
		Phaseolus vulgaris, L.	Para
		Tephrosia purpurea, Pers	Dia
		Alsicarpus monilifer,Dc	Para
		Rhynchosia minima, Dc	Dia
11	Malvaceae	*Malva parviflora*, L.	Dia

Contd...

Table 4.1. Contd...

Sl. No.	Family	Plant Species	Type of Heliotropism
		Malvasturn rotundifolia, A.Gray	Dia
		Sphaeralcea, A.St.Hill	Dia
		Sida acuta, Burn	Dia
		Sida cordifolia, L.	Dia
		Hibiscus cannabinus, L.	Dia
12	Martyniaceae	*Proboscidea parviflora*,	Dia
13	Nyctaginaceae	*Abronia villosa*, Juss	Dia
		Alionia incarnata	Dia
		Boerhaavia diffusa, L.	Dia
14	Portulaceae	*Portulaca oleracea*, L.	Dia
15	Rhamnaceae	*Zizyphus mauritiana*	Para
16	Solanaceae	*Physalis minima*, L.	Dia
17	Sterculiaceae	*Waltheria indica*, L.	Dia
		Melochia corchorifolia, L.	Dia
18	Tiliaceae	*Corchorus acutangulus*, Lam	Para
19	Tropaeolaceae	*Tropaeolum*, L.	Dia
20	Zygophyllaceae	*Kallstroemia grandiflora*, Scop.	Para
		Tribulus terrestris, L.	

diaheliotropic response is known to maximize interception of PAR throughout the day and improves daily total photosynthesis. Paraheliotropism protects leaves from higher light levels and high temperature-induced damage to PS II during midday, while diaheliotropic plants exhibit full protection against these adversaries (Kao and Forseth, 1992; Sailaja and Das, 1996a). These plants do not allocate large resources to either energy-dissipating mechanisms or for repair of damaged photosynthetic apparatus, which are otherwise essential in static plants to overcome photoinhibition. The diaheliotropic plants, exhibit high photosystem II efficiency even under maximal photon energy capture. This unique behaviour was observed in solar tracking plant (diaheliotropic) species with a characteristic ability to maintain constancy in photochemical efficiency and photosynthetic rates throughout the day (Sailaja and Das, 1996a). Using the parameter of Chl a fluorescence, F_v/F_m which is a measure of PS II efficiency (Krause, 1994; Barth and Krause, 2002), has been studied in several plants during different times of the day and was shown to be constant in the diaheliotropic plants (Sailaja and Das, 1996a). Thus, it was demonstrated that under field conditions, diaheliotropic plants lack photoinhibition. The mechanism of overcoming photoinhibition in these plants is not apparent. The construction of photosynthetic apparatus suitable for high light conditions is apparently not the usual evolutionary strategy (Ort, 2001). Therefore, the diaheliotropic plants probably have an inherent as yet unknown strategy for the achievement of maximal photosynthesis under high irradiance. Perhaps the maintenance of relatively higher rates of photosynthesis in the diaheliotropic plants might provide a larger sink for the utilization of the maximal light intercepted by these plants. Photodamage has been avoided by these plants and, therefore, may be studied in greater detail in the future in an effort to understand the improved photosynthetic productivity.

The laser induced F690/F735 ratios remained unaffected diurnally in the diaheliotropic plants, indicating that the quantum yield of photosynthesis and photosystem activity of PS I and PS II were not lowered in diaheliotropic plants during midday (Sailaja et al., 1997). The laser induced F 690/F 735 ratio as a non-destructive method for stress detection is well documented (Chappele et al., 1984; Hak et al., 1990; Lichtenthaler et al., 1990; Agati et al., 1995). An increase in thermal energy dissipation is not noticed (Sailaja et al., 1997). A midday depression in photosynthesis and quantum efficiency has been shown in non-tracking plants (Deming–Adams et al. 1989). The detailed study of the structural components and the functional attributes of PS II to identify the novel mechanisms of photoprotection exhibited by the diaheliotropic plants are very much warranted for future research.

Solar tracking plants exhibiting diaheliotropism displayed a characteristic ability in maintaining constancy in photochemical efficiency and photosynthetic rates throughout the day, in contrast to plants with static leaf-orientation. A higher photosynthetic efficiency of diaheliotropic plants is proposed to be due to greater quantum efficiency of PS II at high irradiances, which would otherwise be detrimental to the paraheliotropic and static leaf plants (Sailaja and Das, 1996a). The authors (Sailaja and Das, 1996a) claimed that this is the first demonstration of lack of photoinhibition under field conditions in solar tracking plants that maximize light interception.

It has also been shown by Sailaja et al., (1997) that an increased thermal energy dissipation was evident in compass plants, while such phenomenon was absent in diaheliotropic plants. The authors have concluded that the findings indicated a lack of down regulation of either PS II or PS I in diaheliotropic leaves which possess the capacity of avoiding high light stress and with gainful strategy for maximization of light energy and photosynthesis.

Horton (unpublished data) working with South American genotypes of bean have shown clear morphological differences among the four selected varieties. The four varieties showed an increased photosynthetic capacity under stress conditions, indicating an acclimation of photosynthesis. All the varieties exhibited paraheliotropic leaf movements and it was concluded that these leaf movements are an important part to abiotic stress. The Fv/Fm, a measure of quantum efficiency of PS II, had only a slight decrease, exhibiting a mild photoinhibition. This situation is ascribed to the occurrence of leaf movements (paraheliotropic), which greatly moderates the amount of light being absorbed and removes the cause of inhibition. The author has observed a genetic variability, even in the selection of just four varieties in the response of leaf movement which perhaps gives an opportunity for genetic analysis and breeding. This is evidenced by stress–sensitive varieties with lower threshold for leaf movements than those of resistant varieties.

The studies on foliar heliotropism were extended to several leguminous and non-leguminous plants by Rao and Das (unpublished) using the parameter, Cos i during the day. Typical results are shown in Table 4.2.

Table 4.2. Diurnal courses of Cosine of angle of incidence (Cos i) of some heliotropic plants. (From Rao and Das (unpublished))

Plant Species	Time (h) of the Day					
	8.00	10.00	12.00	14.00	16.00	17.00
Diaheliotropism-High Fidelity :						
Cleome gynandra, L.	0.95	0.97	0.99	0.97	0.94	0.95
Ricinus communis, L	0.87	0.92	0.92	0.93	0.91	0.88
Euphorbia Hirta, L.	0.88	0.92	0.94	0.92	0.93	0.89
Cyamopsis tetragonoloba, Taub.	0.86	0.87	0.90	0.89	0.86	0.85
Cicer arietinum, L.	0.91	0.95	0.94	0.92	0.91	0.92
Diaheliotropism(Less intense)						
Hibiscus cannabinus, L.	0.91	0.92	0.84	0.84	0.86	0.88
Cajanus cajan, Miisp	0.92	0.91	0.87	0.87	0.86	0.85
Trifolium dubium, Sibth	0.92	0.85	0.77	0.73	0.84	0.94
Corchorus acutangulus, Lam.	0.92	0.83	0.73	0.80	0.87	0.92
Paraheliotropism :						
Phaseolous vulgaris, L.	0.71	0.56	0.52	0.67	0.68	0.68
Vigna unguiculata, (L.) Walp.	0.71	0.66	0.54	0.63	0.63	0.63
Vigna mungo, (L.) Hepper	0.76	0.56	0.56	0.69	0.70	0.76

Values are average of 10 independent readings

The diurnal courses of cos i in the heliotropic plants were quantified through the measurement of cos i. Cosine of indicence (cos i), is a measure of the leaf laminar orientation towards the direct solar beam. Therefore, it gives the actual proportion of the direct beam incidence on the leaf. Cos i is determined as the ratio of the difference between total and diffused photon flux on a leaf/leaflet over the difference between the photon flux on a solar perpendicular and the diffuse value. Accordingly, a Cos i value of 1.0 indicates that the leaf lamina faces pendicular to the direct beam and a value of zero represents a leaf positioning parallel to the solar beam (Kao and Tsai, 1998). The photosystem II resistance for photoinhibition is reconfirmed in solar tracking plants by this study with plants held under natural habitats in open fields. The diaheliotropic plants with high fidelity for solar tracking exhibited cos i values above 0.90 between 0.800 to 1700 hrs of the day, thereby showing a remarkable correlation between leaf azimuth and solar azimuth. It was also observed (Table 4.2) that the paraheliotropic plants show a lowered diaheliotropic behaviour in the forenoon and afternoon, while at midday, they exhibited the typical paraheliotropism. The obligate diaheliotropic plants exhibited high Cos i values throughout the day without midday depression. The remarkable observation was that these diaheliotropic plants did not exhibit avoidance of light interception even under lowering of soil moisture (Rao and Das, unpublished) and, accordingly, it was concluded that perhaps the trait of leaf orientation is genetically fixed rather than environmentally controlled. The belief that the leaf orientation is presumably a genetically-controlled behaviour, finds similarity in the observations of Horton (unpublished) in his studies on bean genotypes. These results warrant a more detailed future investigation on the structural components and function of PS II in the heliotropic plants to unravel the photoptrotective strategies involved.

4.5. SITE OF PERCEPTION AND MECHANISM OF LEAF MOVEMENTS

Two categories of plants exhibiting heliotropic, turgor–mediated, leaf movements are known (Koller, 2000). Plants in one category exhibit laminar phototropism. The direction of sunlight is perceived in their laminae; it has vectorial excitation and tracks the position of the sun. The resulting orientation is diaheliotropic. Plants in another category exhibit pulvinar phototropism. They perceive the direction of sunlight in their pulvinus. Their laminar orientation with respect to the sun changes throughout the day, resulting in changes of interception of light by the pulvinus (Koller and Ritter, 1994). It is the passive outcome of pulvinar phototropism and is not diaheliotropic.

The phenomenon of solar tracking is known from the early studies of Yin (1938) and that it is a blue light response, while the actual photoreceptors involved are not identified (Batschauer 1998; Briggs and Christie, 2002). In view of the fact that solar tracking behaviour results from light-driven turgor changes in the volume of the cells just as in the case of stomatal guard cells, the phototropins might be involved here also (Briggs and Christie, 2002). Sakamoto and Briggs (2002) have suggested that solar tracking may involve phototropins.

The precise molecular mechanisms concerning the heliotropic movements are not currently known, while some knowledge is available on the mechanism of nyctinastic movements. However, Cronlund and Forseth (1995) have studied the mechanism of soybean leaflet movement and concluded that the mechanism of heliotropic movement was similar to that of nyctinastic movements. These authors have studied the role of K+ channels and the plasma membrane H^+/ATPase (Michelet and Boutry, 1995) in paraheliotropic movements through the measurements of leaf movements after treatment of pulvinus with promoters and inhibitors of H^+ATPase and K+ channels. H^+ATPase inhibition reduced the leaf movement by elimination of turgor gradient. The K-channel blocker, TEA (Tetraethyl Ammonium Chloride) also reduce the leaf movement. These results implicate that K^+ Channels work as gates for ion fluxes resulting in turgar–mediated leaf movements.

There is presently only a scattered literature available on leaf orientation in relation to the regulation of light interception. This is an area of considerable interest not only in the understanding of photoprotective strategies, but also to unravel the factors responsible for the inherent superior photosynthetic capacity of diaheliotropic plants which apparently lack photoinhibition and avoid light stress.

5
Acclimation of Photosynthesis to Light Environment

5.1. INTRODUCTION

Acclimation to changing light intensity in the growth environment is essential for the survival of photosynthetic organisms. It involves a relatively long-term adjustment of the components of photosynthetic apparatus to the ambient light environment (Rosenqvist, 2001). In order to mitigate an imbalance in the energy supply and consumption, the process of acclimation obviously involves changes in structure and function of photosynthetic apparatus (Boardman, 1977; Hihara, 2001). The adjustment of the photosynthetic apparatus to the prevailing light regime is referred to as light acclimation of photosynthesis (Anderson et al., 1995; Murchie et al., (2002). Acclimation is specifically a phenotypic expression of the adjustment by the organisms to the growth environment within the limits of the genome (Levy and Gnatt, 1988; Falkowskii and Laroche, 1991; Sailaja and Das, 1995).

Higher plants possess a substantial flexibility for optimal performance under varied light environments (Andersson et al. 2001). This phenotypic flexibility is also called plasticity and is species specific in nature (Evans, 1996). The tolerance of plants for light stress, specially light quantity, is brought about by a dynamic acclimation of photosynthetic apparatus, leading to optimization of the photosynthetic efficiency and the utilization of the imbalanced resources (Anderson et al. 1995). Thus the responses resulting in photosynthetic acclimation to irradiance levels is often specific to a given species (Andersson and Osmond, 1987; Bailey et al., 2001) and are even related to particular habitats (Murchie and Horton, 1998). The strategies adapted by the leaves for maximization of photosynthesis to different light environments are extremely important for an understanding of the behaviour and acclimation to an ever-changing light regime in nature.

5.2. REGULATION

In respect of the regulatory mechanisms responsible for photosynthetic acclimation to the growth light regime, different views are expressed based on specific observations on higher plants. One of the views regards the acclimation as a generalized regulatory mechanism for modulating the photosynthetic apparatus in plants (Melis, et al., 1985) and in cyanobacteria (Fujita et al. 1987). According to this model, the primary signal could be a redox state of cytochrome f or PQ pool, while Chow et al., (1990) and Kim et al., (1999) considered photosystems themselves as the primary sensors. On the contrary, Anderson et al., (1995) have stated that multiple mechanisms mediate the signal transduction, unlike the above suggested model of common regulatory pathway. Under static conditions, the sun and shade plants have specific strategies of adjustment to light intensity. The low light plants (shade plants) possess higher Chl b, LHC II and also higher levels of PS I for maximizing light harvesting. On the contrary, the high light plants have higher cytochrome b6f complex, plastoquinone, plastocyanin and the carbon fixation enzymes (Andersson and Osmond, 1987; Andersson et al., 2001). Chl a/b is an indicator of acclimation to light intensity, the ratio decreasing in low light and with increased values under high light. This suggests a much larger size of light harvesting antenna in low light (Andersson et al., 1997). Differences also exist in the stoichiometry of photosystems. Thus, the low light plants have reduced ratios of PS II/PS I due to fews PS II units of larger size antennae, while the reverse is true for high light plants.

Chloroplast acclimation is affected by a host of environmental factors of which the responses to variations in growth irradiance has been studied quite extensively. Electron transport operates with a rate restriction occurring at PQH_2/Cytochrome b6f under non-limiting light (Andersson et al. 1997b). Large variations in the relative amounts at stoichiometries of the major thylakoid components were detected in relation to changes in growth irradiance. With increasing irradiance, the change observed in the thylakoid organization is a reduction in the number of chlorophylls per electron transport chain. The major change is in the allocation of material resources between light harvesting and electron transport components. The large changes in Cytochrome b6f concentration occur over the lowest irradiances. The plastoquinone pool also increases with increasing irradiance. The total number of PS II reaction centres generally increases with increasing growth irradiance (Genty and Harbinsson, 1996).

At low growth irradiance, a decreased chlorophyll a/b ratio is observed. This is supposed to result from an increase in the absorbance cross-section of PS II relation to PS I content is constant and PS II content decreases with decreasing growth irradiance (Boardman, 1977; Anderson,

1986). The changes are correlated with two parameters, P_{max} and Chl a/b ratio, respectively. A quick assessment of acclimation is determined through these parameters.

Generally, the process of acclimation is regarded to comprise two categories based on the time scale of response (Andersson, 1986; Andersson et al., 1997; Hihara, 2001). The short-term response is completed in a time scale of seconds to few hours and comprises the processes like state transitions, energy dissipation, efficiency of energy transfer from LHC to PS II RC and formation of non-functional PS II RC (Anderson et al. 1997). On the other hand, a long-term response to changing light intensity is truly an acclimation process, as explained above, involving the modulation of structural composition and function of chloroplast and takes a much longer duration — usually several hours to days. Accordingly, the long-term acclimation is brought about either by changes in the synthesis and or degradation of specific components of chloroplasts, while the short-term responses are only due to alterations in the existing chloroplast components (Bailey et al. 2001).

5.3. ACCLIMATION TO IRRADIANCE LEVELS

Quite a number of studies on the acclimation process were related to the characterestic differences between the sun and shade plants representing the high and low light regimes (Boardman, 1977; Anderson, 1986; Anderson and Osmond, 1987; Thayer and Bjorkman 1990). However, in recent years, studies conducted on acclimation have emerged as an active area of photosynthesis research by documenting responses of several species of higher plants and also of algae under fluctuating levels of irradiance.

Several plant species have been investigated in respect of photosynthetic acclimation under different levels of irradiance exclusively. Changes in the ratio of chl a/b such as increased values under high light conditions compared to low levels of irradiance in pea (Demmig – Adams and Adams, 1996a; Leong and Andersson, 1984), barley (De la Torre and Burkey 1990), *Arabidopsis* (Walters and Horton 1994), in spinach (Lindahl et al., 1995) and in *Amaranthus* (Sailaja and Das 1995) have reported such findings. Figure 5.1. shows the acclimatory changes in chlorophyll a/b ratio in *Amaranthus* (Sailaja and Das, 1995). It was also observed that the size of PS II light-harvesting antennae decreased with increased light intensity. Further, changes in electron transport system have also been noticed. Pronounced increases in the cytochrome b6F complex under higher irradiance have been reported for several species. Cytochrome f content was much higher in the leaves of plants grown under high irradiance compared to that in leaves of low irradiance plants (Evans,

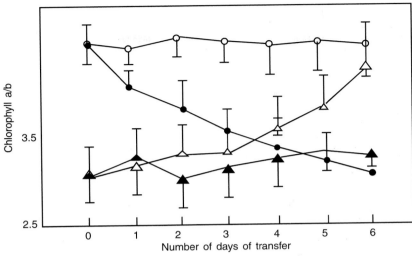

Fig. 5.1. Time course of acclimation determined by Chl a/b ratio to an altered light regime in mature leaves. After transfer of plants from high to low light and vice-versa, the intial values of Chl a/b are restored in six days, thus demonstrating the protential of leaves for acclimation to varying irradiance levels. Open circles – static high light intensity (2000 µEm^{-2}s^{-1}) solid triangles – static low light level (650 µ Em^{-2}E^{-1}) solid circles- plants transferred from high to low light. Open triangles- plants transferred from low to high light. (Figure reprinted with permission by Kluwer Academic Publishers from M.V. Sailaja and V.S. Rama Das (1995): Photosystem II acclimation to limiting growth light in fully developed leaves of *Amaranthus hypochondriacus* L; an NAD – ME C$_4$ plant. Photosynth. Res. **46**: 227-233. Copy Right 1995, Kluwer Academic Publishers)

1996; Reddy et al., 1983). Stoichiometry of reaction centres has also been shown to fluctuate in relation to irradiance levels. PS II content has increased under high light in pea (Leong and Andersson, 1984; Evans, 1996), mustard (Wild et al., 1986) and in *Arabidopsis* (Walters et al. 1999). It is, therefore, clear from the above findings that in general, the acclimation process involves alterations in Chl a/b ratio, cytochrome b6f complex and the reaction centre stoichiometry.

Acclimation, specifically to low growth irradiance or in other words, light limiting conditions, was investigated in *Amaranthus hypochondriacus* by Sailaja and Das (1995, 1996b). Wherein it was observed that in the bundle sheath chloroplasts, a fixed number of PS I reaction centers exists while PS II centers were reduced. The coordination between the photosystems was brought about by lowered levels of cytochrome b6f complex (Sailaja and Das, 1996b). These researchers have further extended this study to acclimation strategies to two other C4 species, *Eleucine coracana* and *Gomphrena globosa* (Sailaja and Das, 2000). These studies were also carried out with a perspective of time course (duration of acclimation) process in the C4 plants, while studies on C3 plants, *Solanum, Pisum, Lycopersicon, Phaseolus* and *Hordeum* were conducted by others earlier (Von Caemmerer and Farquhar, 1984; Davis et al., 1986; Ferrar and Osmond, 1986; Chow and Anderson, 1987; DeLa Torre and Burkey, 1990).

5.4. ACCLIMATION IN MATURE LEAVES

Acclimation of PS II to limiting growth light in fully-developed leaves was investigated in *Amaranthus hypochondriacus*. This study has demonstrated that acclimation is not necessarily restricted to a leaf development process but the fully-developed leaves exhibit capacity to acclimate to a new growth light regime. The Chl a/b ratio is a measure of relative proportion of RC to that of LHC. A decrease in Chl a/b ratio under reduced irradiances is to be thought of as an increase in LHC at the expense of RC. In this context, recently, Oguchi et al., (2003) have studied in detail the light acclimation of photosynthesis in mature leaves of *Chenopodium album* and demonstrated that leaves transferred from low growth light to high light had increased light-saturated rate of photosynthesis (Pmax). This transfer has also resulted in an increase in the area of chloroplasts facing intercellular spaces. The leaf thickness of transferred plants has not increased to the level of high light grown plants, therefore, determining the upper limit of Pmax. It is already known that photosynthetic capacity is correlated with leaf thickness (Jurik, 1986; Hanba et al., 2002). The work of Oguchi et al., (2003) has shown that chl a/b ratio increased in leaves when transferred to high light, thus, implying a change in the physiology of chloroplasts and also in the size of chloroplasts.

It is apparent that the existence of high plasticity in photosynthetic light acclimation of mature leaves is of great signifinance and beneficial under conditions of sudden increased growth irradiance (Kursar and Coley, 1999; Yamashita, 2000; Oguchi et al., 2003). The increase of Pmax when the leaves were transferred to high light in *Chenopodium album* was attributed to an increase in the surface area of chloroplasts facing intercellular space.

From the study made by Sailaja and Das (1995), it was concluded that under suboptimal light conditions, there is an impairment of functional open PS II RC. The changes in quantum efficiency of PS II (Fv/Fm) and the regulation of oxygen evolution under limiting light are shown in Figures 5.2 and 5.3, respectively. The impairment of functional open PS II reaction centres under suboptimal light conditions has been shown by the reduced Fv/Fm ratio. A subsequent increase in the Fv/Fm ratio showed that adjustment to the lowered light conditions takes place for the improvement of energy transfer mechanisms (Figure 5.2). The pattern of oxygen evolution (Figure 5.3) also demonstrated the acclimatory response over a period of six days. The strategy for acclimation to low light (650 μ E m^{-2}s^{-1}) was a two-pronged response. One of those responses was a modulation of the size of LHC II, while the other response involves a decreased synthesis of RC protein D1, due to depressed transcription of psbA gene. Accordingly, the molecular regulation to low light stress in *Amaranthus hypochondriacus* is achieved by interaction of an increased antennae size, reduced D1 protein synthesis and through decreased functional RC. The fact that the LHC of PS II has preferentially increased compared to RC was also demonstrated by an increased F690/F735 ratio. The authors have concluded that the strategy for acclimation of PS II function under limiting light environment involves reduced synthesis of RC components, thus, minimizing the costs of construction. Again, the finding that the psb A gene is transciptionally regulated in *Dunaliella* growing under low light regime has been shown earlier. Therefore, the low light stress in a fully-developed higher plant system, appears to be regulated according to the above studies by Sailaja and Das, (1995) through an interaction of increased antenna size, reduced D1 protein synthesis and decreased functional RC.

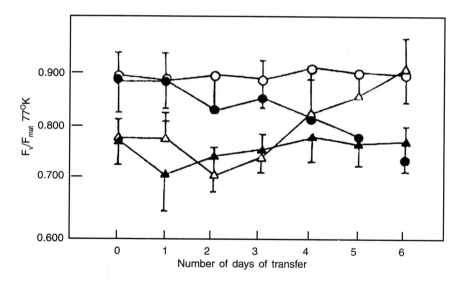

Fig. 5.2. Acclimatory response of mature leaves to quantum efficiency of PS II (Fv/Fm) on transfer to altered light regime. (Figure reprinted with permission by Kluwer Academic Publishers from M.V. Sailaja and V.S. Rama Das (1995): Photosystem II acclimation to limiting growth light in fully developed leaves of Amaranthus hypochondriacus L; an NAD – ME C$_4$ plant. Photosynth. Res. **46**: 227-233. Copy Right 1995, Kluwer Academic Publishers)

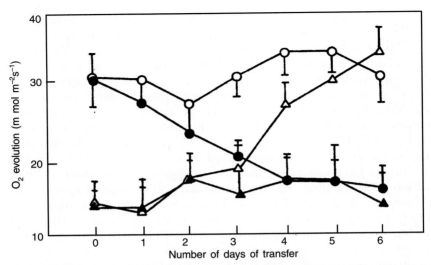

Fig. 5.3. Photosynthetic acclimation of mature leaves to changed light environment as determined by pattern of oxygen evolution. Open circles – Static high light intensity (2000 $\mu Em^{-2}s^{-1}$) Solid triangles – static high light level (650 $\mu Em^{-2} s^{-1}$) solid circles- plants transferred from high to low light. Open triangles- plants transferred from low to high light. (Figure reprinted with permission by Kluwer Academic Publishers from M.V. Sailaja and V.S. Rama Das (1995). Photosystem II acclimation to limiting growth light in fully developed leaves of Amaranthus hypochondriacus L; an NAD – ME C_4 plant. Photosynth. Res. 46: 227-233. Copyright 1995, Kluwer Academic Publishers)

Acclimation to reduced growth irradiance in *Eleusine coracana* (NAD-ME) and *Gomphrena globosa* (NADP-ME) was investigated (Sailaja and Das, 2000). A superior photosynthetic acclimation to reduced irradiance in *G. globosa* was due to a smaller change in the functions of cytochrome b6f, PS I and PS II, leading to higher photosynthesis than in *E.coracana*. The PS II acclimation to limiting light has also been investigated in relation to C4 metabolism. The plants of different C4 sub-types showed differential acclimation patterns to reduced irradiance. The size of LHC II was higher in NAD-ME C4 plant, *E.coracana* compared to NADP-ME C4 plant, *G.globosa*. This is to overcome a large decrease in the efficiency of excitation energy capture by open reaction centres (Sailaja and Das, 2000). Also, a smaller change in the cytochrome b6/f complex, in NADP-ME compared to NAD-ME type is related to a lower decline in electron transport rate and functional PS II RC. These authors (Sailaja and Das, 2000) have identified two modes of acclimation to reduced irradiances : (1) *G.globosa* type which combines modulation of composition and function of thylakoid membrane to higher performance; and (2) *E.coracana* type, where a greater investment in light harvesting apparatus was noticed to achieve a reasonable acclimation level.

Amaranthus, an NAD-ME C4 dicotyledonous plant, showed acclimation similar to *E.coracana* type which is also an NAD-ME monocotyledonous species. The NADP-ME type have superior mode of acclimation compared to NAD-ME. It is significant that NADP-ME is more evolved compared to NAD-ME. It was also observed (Sailaja and Das, 1996) that certain responses to cell-specific modulation occurred in the mesophyll and bundle sheath chloroplasts. In the latter the situation appears to be a strategy of fixed number of PS I reaction centres, presumably related to the requirement of reduced PS II reaction centres. The lowered level of cytochrome b6f complex is perhaps responsible for the coordination between the photosystems in the bundle sheath. Contrary to this, in the mesophyll chloroplast, a reduction in both the functional photosystems was the strategy for acclimation to reduced irradiances. (Sailaja and Das, 2000). These differences in the mode of acclimation require further investigation.

5.5. ACCLIMATION TO CHANGING LIGHT REGIME

Considerable work in recent years has been done on the acclimation responses to changing light environment in *Arabidopsis thaliana* (Walters et al., 1999; Walters and Horton, 1994, 1995; Yin and Johnson 2000; Weston et al., 2000; Bailey et al., 2001). Walters et al., (1999) have investigated the chloroplast composition in mutants of *Arabidopsis thaliana* grown under both high and low irradiance and observed that photoreceptors do not have a major role in the regulation of photosynthetic acclimation. Bailey et al., (2001) have monitored the process of photosynthetic acclimation in six growth irradiance levels ranging from 35 to 600 µmol quanta $m^{-2} s^{-1}$ using a single plant species *Arabidopsis thaliana*. The responses to changing light environment may take place at the level of leaf and or the chloroplast. The leaf level responses include changes in the numbers of chloroplast per unit leaf area and leaf thickness. These authors believe that the capacity for photosynthetic acclimation is species specific. The regulatory mechanisms underlined the acclimation response to irradiance are not well defined. However, a common notion appears to be a self regulatory feed back, sensing imbalance in photosynthesis which leads to adjustments in the chloroplast composition for correction of that imbalance.

The authors (Bailey et al., 2001) have stated that a multiple regulatory mechanism exists for photosynthetic acclimation and have demonstrated in *A. thaliana*, the existence of separate responses for growth at low and high irradiances. Two distinct strategies of acclimation at different light intensities have been proposed based on differential changes in the major LHC of PS II, shifts in the content of both RCs and also highly significant changes in the levels of minor LHC II and LHC I components. In the low light acclimation (35 µmole m^{-2} s^{-1}), a striking feature was a twofold increase in PS I content but this increase is not accompanied by increase in any of LHC I polypeptides. In contrast, in addition to dramatic increases in the bulk LHC II components, Lhcb1 and Lhcb2, there was also a twofold increase in the levels of minor LHC II components, Lhcb 5 (CP 26) and L hcb6 (CP 24) though there is a decline in PS II levels. Acclimation to low growth light involves regulation of nuclear-encoded Lhcb1 and Lhcb2 gene expression resulting in an increase of LHC II content in plants has been reported earlier (Green and Salter, 1996). Growth at high irradiance includes a dramatic loss of LHC II with decreases in the content of Lhcb 1 and Lhcb 2. PS II levels increase in high light and also a twofold decline in the minor LHC II polypeptides thus, a decrease in PS II antennae size offers resistance to photoinhibition.

Lhca4 content decreased under high light (600 µmole m^{-2} s^{-1}). The decrease in Lhca4 in high light is much higher than that of Lhca1. Bailey et al., (2001) believe that photosynthetic acclimation at different light levels is quite complex and perhaps distinct strategies for acclimation at high and low light may be envisaged. Maxwell et al., (1999) have investigated chloroplast acclimation to high light in an epiphyte *Guzmania monostachia*. Acclimation has resulted in a reduction in the size of chloroplast, thylakoid volume and granal stacking is a reduction in amounts of LHC I, LHC II and OEC 33. Maxwell et al., (1999) have concluded that the acclimation strategy in this epiphyte is different from the conventional strategies possesed by other plants. Therefore, this appears to be a unique strategy confined to the epiphytic bromeliads consistent with the epiphytic habitat.

Yin and Johnson, (2000) have investigated the capacity for acclimating to fluctuating light regime, for varying times between few minutes to three hours, using three plant species *Arabidopsis thaliana, Digitalis purpurea* and *Silene dioica*. The plants grown at a constant light of 100 µmols m^{-2} s^{-1} were transferred after six weeks to light intensities of 475 or 810 µmol m^{-2} S^{-1} for seven days. It was observed that the acclimation depended on the species concerned. This study was claimed to have documented for the first time the ability of fully developed leaves to acclimate to a new growth light environment. However, as mentioned above, Sailaja and Das, (1995) have already demonstrated the capacity of fully-developed leaves for acclimation using *Amaranthus hypochondriacus*. Yin and Johnson (2000) have observed that an intermittent high light signal is adequate for triggering the acclimation process. *D.purpurea* had possessed a lower response compared to the other two species. The characteristic response to changing irradiance levels were related to the content of rubisco and cytochrome f. When the irradiance was increased upto 475 µmol m^{-2} s^{-1}, an increase in rubisco content was noted with little change in cytochrome

f. The pattern reversed between irrdiances 475 to 810 µmol m^{-2} s^{-1} with a definite increase in cytochrome f and no change in rubisco. In fact, such a relationship between growth light intensity and cytochrome f content was shown by Reddy et al., (1983) in *Sorghum* cultivars, where the high levels of cytochrome f were associated with high light intensity between 900 and 1800 µE m^{-2} s^{-1}, while it decreased in low light levels. The study made by Yin and Johnson (2000) has shown that there was a lag period of one to two days before a noticeable change in acclimation occurred, which was interpreted to be due to the delayed signal transduction pathway. It is well established that the acclimation of plants transferred to high light from low growth light environments is brought about by lowering of the antennae size of the photosystem II. This reduction of antenna size is presumably brought about by an acclimative proteotysis of LHC II. Yang et al., (1998) using spinach plants grown at low light intensity (30 µE m^{-2} s^{-1}), have transferred to a higher irrdiance level of 600 to 800 µE m^{-2} s^{-1}, and investigated the process of acclimative proteolysis of LHC II. The authors observed a delay of two days after transfer to high light per the acclimatory response to be noticed. In this respect, the occurrence of delayed acclimative response in this species in similar to such an observation of response to acclimation in *Arabidopsis* and other species (Yin and Johnson, 2000). The proteolysis was found to be ATP-dependant and is of serine or cystine type and located in stroma-exposed thylakoids, while the identity of LHC II protease could not be established. It is different from other ATP-dependant proteases, CLP and FtsH which are concerned with D1 protein degradation. Subsequently, Yang et al., (2000) have extended this work to further understand the substrate specificity and the recognition site for the protease. They have shown that several steps in the reduction of LHC II under increased irradiance which involves changes in oligomeric state, phosphorylation and lateral migration are perhaps controlled through the N-terminal region of LHC II. Their findings have established that the crucial importance of N- terminus for the LHC II degradation.

Besides the higher plant systems for studies on photosynthetic acclimation, investigations were also conducted in algal systems, particularly *Chlorella* (Wilson and Huner, 2000), *Synechosystis* (Hihara et al., 2001), *Dunaliella* (Escoubas, et al., 1995; Kim and Melis, 1992; Masuda et al., 2002), and *Dunaliella* and *Phaeodactylum* (Quigg et al., 2003). Wilson and Huner (2000) have made a detailed study of long-term photoacclimation to growth irradiance in *Chlorella*. They have demonstrated that the changes involved in photoacclimation are regulated by a redox state of plastoquinone pool and or cytochrome b6f complex. This finding is obviously similar to that of Yang et al., (2001) in a higher plant, *Lemna*. Yang et al., (2001) have examined the role of cytochrome b6/f in the photoacclimation process in *Lemna perpusilla*. On the basis of their studies, they have arrived at the conclusion that plastoquinone redox level is the major signal in regulating the LHC II photoacclimation in plants exposed to high or limiting light conditions. Cytochrome b6/f was considered to play a secondary role based on their studies on *Lemna*.

PS II/PS I ratio is known to increase when cyanobacteria are exposed to high light (Murakami and Fujita, 1991; Hihara et al., 1998). A decreasing PS I content is believed to confer protection to high light damage. Recent studies by Hihara et al., (2001) have established the fact that the transcriptional levels of genes responsible for encoding the subunits of photosystems are involved under changing light ocnditions in bringing about photosystem stoichometry. It was shown that psa genes were repressed to a greater extent than that of psb genes when exposed to high light. Hihara et al., (2001) have made a detailed DNA microarray analysis of cyanobacterial gene expression using *Synechocystis*, during acclimation from low to high light intensity. A pattern of gene expression during acclimation to changing light intensity has been provided. Around 160 genes were identified and classified. Thus, the authors have provided an evidence for the expression of the genes responsible for acclimative changes to varying irradiances.

The studies on the ability of leaves for photosynthetic acclimation to light environment in the individual plants has been discussed above. However, the leaves in the canopies also possess the capacity for acclimation to light environment (Jarvis et al., 1976; Jarvis and Sandford, 1986). A relationship between the amount of light received by leaves situated in different levels of canopy and the photosynthetic rate has been studied. (Jarvis and Sandford, 1986). The Lower leaves in a canopy possess much lower light saturated rate of photosynthesis than those leaves that receive higher irradiance in the upper part of the canopy. Recently, Meir et al., (2002) have made a detailed investigation of the acclimative responses to irradiance in the photosynthetic capacity of leaves in tree canopies in four different forest trees. The most important finding was that acclimation to irradiance occurred in all the canopies; however, the acclimation was only partial.

It would be of interest to examine the photosynthetic acclimation to cold by overwintering evergreens (Huner et al. 2003). The evergreen conifers avoid photo-oxidation and high light stress during winter months when the rate of CO_2 assimilation is much reduced due to low temperatures by a specific strategy. The winter foliage possessed reduced antennae size, partial loss of PS II and also the development of sustained NPQ for dissipating excess light as heat. Major alterations in the composition and organization of PS II antennae were observed as an acclimative response for protection against excess light absorbed during winter (Huner et al. 2003). The differences between the leaf strategies of deciduous and evergreens have been also discussed by Huner et al. (2003).

The fact that the ability of acclimation can also be systemic and acquired has been shown recently by Karpinski et al. (1999). The authors have investigated *Arabidopsis* plants which have previously adapted to low light by exposing them to excess excitation energy (EEE) and observed the occurrence of a systemic acclimation. Accordingly, it is thought that a mechanism exists to communicate EEE so that a defense against stress may develop. The systemic acquired acclimation would consist of redox changes in the PS II, H_2O_2 and induction of antioxidant defenses. When the leaves were treated with H_2O_2 and then exposed to EEE produce, a lesser degree of photo-oxidative stress was evident. The data shows that H_2O_2 is involved in acclimation to conditions produced by EEE. Karipinski et al., (1999) suggest that the systemic signal can promote redox changes in the proximity of PS II in unstressed chloroplasts. This will induce a protective mechanism in remote chloroplast and cells, a behaviour that has been termed as systemic acquired acclimation.

It is apparent that a considerable number of studies of acclimative process of photosynthesis to varied light irradiances are available for a number of species of higher plants as also of algae. However, much remains to be understood in relation to the actual determining factors and the signal transduction.

6
Transgenic and Biotechnological Approaches

6.1. INTRODUCTION

The application of molecular, biological and biotechnological approaches have considerable potential in photosynthetic research for a better understanding of the precise mechanisms involved. The physiological, biochemical and biophysical investigations have yielded substantial information on the responses of plants in photosynthetic performance related to environmental variables, including growth light intensity. Several studies involving molecular biology and transgenic technology are available in recent years that can be applied to environmental stresses (Raines and Lloyd, 1996) as also in the area of photosynthesis. In addition, the recently emergent genomic technologies are also being applied to the study of photosynthesis (Pesaresi et al., 2001; Dent et al., 2001). Particular stress is laid on the functional genomics also employing the reverse genetics approach in understanding the function of photosynthetic genes in *Arabidopsis*, a study that is being vigorously pursued currently (Pesaresi et al., 2001; Dent et al., 2001).

Several examples of investigations that have principally employed the above-mentioned tools of molecular biology and biotechnology are discussed in this chapter, though a reference to some of these has already been made earlier in the relevant chapters. The subject matter of application of these tools is treated under different aspects of photosynthetic functions, including the structure of photosystems, light harvesting antenne, photoinhibition and related areas.

6.2. PHOTOSYSTEMS : REACTION CENTRES

Swiatek et al., (2001) have made an elaborate study of the function of the protein PsbZ which is a product of the gene ycf 9 in *Chlamydomonas* and of ORF 62 in tobacco which is also chloroplast encoded. They have employed a reverse genetic approach and constructed strains where ycf 9 gene was replaced by a selectable marker cassette. The authors have also used the chloroplast transformation technique, a special tool for generating modified plants through chloroplast transformation, resulting in the production of transplastomic plants. In these transformed plants, the ycf 9 gene was replaced with aadA cassette in the antisense orientation.

Swiatek et al., (2001) have established the function of PsbZ in the interaction of PS II core with LHC antenna. This is evident from the finding that the absence of PsbZ in tobacco has led to growth defects and higher light sensitivity. The role of PsbZ could be further understood by its position near the interface PsII – LHC II of PS II core (Zouni et al. 2001). In the model of supra molecular organization of the PS II (Zouni et al., 2001), several TM helices occur adjacent to D1 and CP 43. Perhaps PsbZ is responsible for these helices in closeness to D1, CP 43 and CP 26. Swiatek et al., (2001) have clearly proved the crucial role of PsbZ in the PS II and LHC II interaction and in the formation of NPQ (Eckardt, 2001). Thus, the study conducted by Swiatek et al., (2001) is an example of the use of molecular biological technique, the specifically targeted gene inactivation, in order to confirm the role of different protein components in the participation of the architecture of PS II as also in the energy dissipation process such as NPQ.

It is clear that the overall organization of PS II involves over 17 different subunits. In brief, the reaction centre of PS II is composed of D1 and D2, CP 47 and CP 43, Cytochrome b 559 and several smaller subunits of lesser known function (PsbH, PsbI, PsbK, PsbL, PsbM, PsbX) (Zouni et al., 2001; Eckardt, 2001). PS II exists as a dimer and the dimer forms a PS II – LHC II supercomplex with LHC II also known to exist as a dimer (Shi et al., 2000; Zouni et al., 2001).

Further, the low molecular weight proteins, PsbH, PsbK and PsbL function in the dimer stabilization. However, Shi et al., (2000) have earlier implicated PsbW, a nuclear-encoded protein, in the stabilization of *Arabidopsis* dimeric PS II complex.

Shi et al., (2000) have used the antisense technique in producing transgenic *Arabidopsis* lines, with much reduced levels of PsbW protein, a low molecular weight subunit of PS II. The phenotypes of wild and antisense transformants were not different from each other under normal growth light intensities. The transgenic plants lacking PsbW protein did not exhibit dimerization of PS II supra complex.

The findings of Shi et al., (2000) are very relevant in determining the precise organization of PS II in vivo in more than one way. One of these methods leads to the conclusion that PsbW is essential in the formation and function PS II dimers. There is a strong evidence to suggest that the PS II is a dimer in the natural state based on the isolation of dimeric PS II – LHC II supra complexes from spinach thylakoids. (Eshaghi et al. 1999). PS II, in its monomeric and dimeric forms contains a large number of low molecular mass proteins (< 7 kDa). PsbW, one of these low molecular mass proteins, is nuclear coded with a mass of 6.1 kDa and is conserved in spinach, *Arabidopsis* and *Chlamydomonas*. PsbW protein is a subunit of PS II only. The other implication of these findings relates to the nuclear control of photosynthetic performance by production of components needed for the optimal structure of PS II. This point is obvious in view of the gene for the production of PsbW is nuclear encoded. The results of these authors (Shi et al., 2000) lead to an important conclusion that the nuclear-encoded PsbW protein is an example of how plant cell nucleus controls the photosynthetic activity of the chloroplast. It remains to be established whether PsbW or PsbZ is a more likely candidate in the dimerization of PsII supracomplex.

Although the role of dimeric organization of PS II is not fully understood, the amount of functional PS II is reduced in the absence of dimeric form. The use of the transgenic study to elucidate the structural organization of PS II has been demonstrated by this study of antisense PsbW plants (Shi et al. 2000).

More recently, Varotto et al., (2002) have studied the functions of individual subunits of photosystem I by generating mutants through transposon insertions which were identified through reverse genetic screening technique. While using *Arabidopsis* En insertion lines with single and double knockouts of genes encoding G,K and H subunits of PS I, the functions of these subunits were investigated. The study of Varotto et al., (2002) has confirmed the earlier findings of Haldrup et al., (2000) and Varatto et al., (2002) that in the absence of specific subunits, the quantities of other PS I polypeptides are altered. Therefore, the assignment of precise role for each subunit is not possible. When PS I-G is absent, most of the proteins comprising PS I core are altered or affected in their levels, but not the Lhca proteins. Therefore, PS I-G has a role in stabilizing PS I core. Varotto et al., (2002) have suggested that PS I-K has a definite role in state transitions. Jensen et al., (2002) simultaneously, have also generated antisense lines of *Arabidopsis* for the study of the functions of PS I-G subunit. These authors have transformed the *Arabidopsis* plants by using PsaG cDNA in the antisense orientation. They have observed that PS I-G has a specific role in stabilizing the binding of peripheral antennae to the core. Their data demonstrated that PS I-G has no specific role in the functioning of PS I antennae. Further, it was shown that PS I-G has a considerable role in regulating PS I, since in its absence, the activity of PS I is increased. Under high light regimes, PS I-G has a regulatory role since under high irradiance, the photochemistry of PS I surpasses that of PS II. Ihalainen et al., (2002) have suggested, using pigment analysis and absorption spectra with *Arabidopsis* PS I complexes, that PS I-K and PS I-G are involved in the interaction between core and peripherial antennae proteins. The work of Ihalainen (2002) has demonstrated that PS I-K protein has a similar function to that of PsbZ (Swaitek et al., 2001) of PS II in connecting the peripheral antennae to the core complexes.

Ruffle et al., (2001) have used the algal system, *Chlamydomonas reinhardtii* with site-directed mutations in the Psb A and Psb D genes, for D1 and D2 proteins, respectively. The transformants thus obtained were used for the investigation of the peripheral accessory chlorophylls in photosystem II. Their studies clearly dmonstrated that the mutants, D1 – H118 and D2 – H117 differed substantially in the light harvesting efficiency and sensitivity to photoinhibition. It was shown that D1 – H118Q mutant was more resistant to photoinhibition than the wild type. This could perhaps be resultant of reduced light harvesting. They have also stated that their study with transgenics has shown that the two peripheral accessary chlorophylls have dissimilar functions, directed to linear electron transfer, light harvesting and photoprotection.

6.3. LIGHT HARVESTING ANTENNAE

Ruf et al., (2000) have made a detailed investigation of a structural component coded by the gene ycf 9 and its function in LHC of PS II using transformed tobacco plants. These authors have made chloroplast transformation producing transplastomic tobacco lines. ycf 9 knockout allele was generated and introduced into the chloroplast genome of tobacco plant. Ruf et al., (2000) have used the reverse genetic approach for constructing a knockout allele, leading to the disruption of ycf 9 in tobacco. Their results indicated that ycf 9 gene is a genuine component of the genome. It was clear from the study that the lack of ycf 9 resulted in destabilization of CP 26 and, therefore, they suggested that the ycf 9 knockout plants provide a useful experimental material for understanding the function of this minor LHC component.

Anderson et al., (2001) have demonstrated the importance of the use of antisense inhibition technique in the production of *Arabidopsis* lines deficient in specific proteins. This also confirmed the earlier studies by Zhang et al., (1997) and Ganetag et al., (2001) in the utility of this technique. Andersson et al., (2001) produced antisense constructs for transformation of *Arabidopsis*, resulting in transgenic lines, for low quantities of CP 26 and CP 29. It is a well- known fact that CP 29, CP 26 and CP 24 (Lhcb 4, Lhcb 5 and Lhcb 6 genes) are the minor antennae complexes forming a part of PS II outer antennae (Jaekowski and Jansson, 1998; Jansson, 1999). Also, that excess light is absorbed, leading to stress. To prevent damage safety systems, particularly the process is called qE exits, which is observed as NPQ. CP 29 and CP 26 are involved in QE, in view of the fact that they are located between the inner antennae and major LHC II.

The authors (Andersson et al., 2001) have reported successful production of CP 29 and CP 26 in antisense lines in *Arabidosis*. The study has revealed that both these proteins are involved in light harvesting and they also influence NPQ — probably in an indirect manner.

As shown by Ruf et al., (2000), protein levels of CP 26 were reduced substantially in tobacco with deletion of ORF 62 (ycf 9). However, the results of Andersson et al., (2001) suggest that the phenotype of ORF 62 is not related to loss of CP 26. The data suggested that CP 29 and CP 26 are involved in the coordination of the LHC II antennae and not directly involved in NPQ (Andersson et al., 2001). CP 29 antisense lines also had decreased amounts of CP 24. This study obviously indicated that a plant can complete its full life cycle under static conditions even when they lack one or more LHC proteins. Further, the results also suggest that under highly fluctuating light intensity, all the LHC proteins are required for efficient function. Thus, the above investigations using antisense lines and reverse genetics approaches have given rise to better understanding of the functions of the proteins. However, a much greater detailed study in this direction is to be encouraged.

The light harvesting complex of photosystem I was studied by Ganetag et al., (2001) who have also transformed the *Arabidopsis* plants with an antisense technique to study its functions, directing their attention to Lhc a2 and Lhc a3. In *Arabodopsis,* the genes responsible for the synthesis of LHC I proteins (Lhca1 to 4) are located in the nuclear genome. They vary in size from 20 to 24 kDa. Since the standard forward genetics procedure is inadequate to establish the functions of each of the PSI proteins (at least 17 polypeptides), reverse genetics is a more powerful tool in dissecting the functions of these proteins. Ganetag et al., (2001) have shown two examples of successful reverse genetics by specifically repressing the expression of the two different LHC I proteins and the transgenic lines are used to determine their structure and function. By way of these techniques, it was shown that the absence of Lhca2 and Lhca3 did not affect the xanthophyll cycle. It is also claimed by the authors that the reverse genetics approach is effective in the study of different polypeptides in vivo. It was observed that more LHC II is associated with PS I in the antisense lines, compensating for the loss of LHC I. The antisense plants also showed an increased reduction of PQ pools if PS I antennae size is decreased. Lhca2 and 3 contribute to the long wavelength fluorescence. The work also suggested that these two proteins are in physical contact with each other and this association is required for stability. Zhang et al., (1997) using an antisense technology, have obtained a reduction of expression of Lhca4 in *A. thaliana*. In one of the lines, Lhc a4 was totally absent. A reduction in the level of Lhca4 did not result in the depletion of Lhca1. This has established that although Lhca1 and a4 would normally make up one complex but can also accumulate in the absence of each other.

Flachmann (1997) studied the PS II antennae composition under varying light conditions in tobacco plants transformed with antisense technique. An increase of PS II antenna size was observed under low irradiance and also higher LHC II content. The results also suggested that LHC II biogenesis is perhaps not controlled by transcription.

The foregone account of different studies using transgenics have inmmensely helped by adding new dimension in our understanding of the structure and function of the photosystem core complexes and of the antennae systems related to both PS II and PS I.

6.4. PHOTOINHIBITION : TRANSGENICS

A fairly large number of studies have also been directed using transgenic technology to understand the process of photoinhibition.

Tyystjarvi et al., (1999b) have made a study of photoinhibition of PS II in tobacco and poplar plants. The tobacco cultivars were expressed with bacterial *gor* gene in the cytosol and Fe SOD gene from *Arabidopsis thaliana* rather in the chloroplast. The transformations were affected as an overexpression of glutathione reductase in tobacco and superoxide dismutase in poplar. This transformation resulted in the activities of glutathione reductase in tobacco leaves and superoxide dismutase in poplars were five to eight times higher than in the untransformed plants. The experiments of the authors (Tyystjarvi et al., (1999b) with the transformed plants have led to some important clues regarding the identity of Active Oxygen Species and the mechanisms. There was a lack of protection by overproduction of SOD in the stroma, suggesting that superoxide is not accessible to dismutation by the stromal enzymes. Protection by glutathione reductase suggested that a soluble reductant has a limited chance to trap the species before it reacts with PS II RC. It was concluded (Tyystjarvi et al., 1999b) that much further work is required to understand the molecular mechanism of loss of PS II activity.

H.Y.Yamamoto and his scholars have made several studies manipulating the levels of the enzymes of the xanthophyll cycle through transgenic techniques. Verhoeven et al., (2001) have investigated the effect of suppression of Z in tobacco plants with an antisense construct of VDE in growth chambers. Under short-term (2 or 3h) high light treatment, antisense plants had a greater reduction in Fv/Fm ratio relative to wild type, which implied a greater susceptibity to photoinhibition. In the long-term highlight stress experiment, the antisense plants had significant reduction in Fv/Fm. The authors concluded that XC-dependent energy dissipiation is critical for photoprotection in tobacco under excess light in the long term.

Sun et al., (2001) have further extended the study of tobacco plants transformed with an antisense construct of VDE and the consequence of reduced XC activity on photoinhibition under field conditions. Antisense plants did not exhibit greater susceptibity to photoinhibition, in such circumstances. Their results indicated that the photoinhibition under field conditions is not dependent on the levels of Z and A. The authors have also stated that the antisense tobacco plants have a differential response to photoinhibition from the one shown by *Arabidopsis* NPQ1 mutants. The lack of photoinhibition under field conditions is presumably related to the characterstics of tobacco plants which are adapted to high light, unlike *Aarbidopsis* which is more of a shady plant.

The data obviously indicated that suppression of A and Z formation affects susceptibility under growth chamber conditions only and not under field conditions. Sun et al., (2001) have also concluded that supressed VDE activity is not critical for growth of sun plants under natural environments and that xanthophyll cycle may be one among the several photoprotective systems, perhaps working in parallel or in series against excess light.

Hieber et al., (2002) have made studies on the overexpression of VDE through transformation of tobacco plants. Since there was a low specific activity with expression in *E. coli*, the *Arabidopsis* VDE was expressed in tobacco. Tobacco, transformed with *Arabidopsis* VDE, had a VDE activity greater by 13 to 19 times than in the wild type. They have further investigated the deletion mutations of C-terminal region of VDE and established that this is important for binding VDE to the thylakoid membrane. The transformed plants showed an increase in NPQ initially with increased rate of de-epoxidation. They stated that further studies with plants overexpressing VDE could lead to a better understanding of the role of xanthophyll cycle in photoprotection. It was suggested that further work in relation to the functional role of overexpressed VDE and xanthophyll cycle in relation to photoprotection is required. Hieber et al., (2002) have stated that it was the first report of overexpression of vilolaxanthin de-epoxidase (VDE). Their earlier work (Sun et al., 2001, Verhoeven et al., 2001) indicated that xanthophyll cycle is important for photoprotection. It is also apparent that the XC (Xanthophyll Cycle) is not an exclusive mechanism for photoprotection.

Grasses et al., (2001) have generated transgenic tobacco plants with reduced CHL P (geranylgeranyl reductase) and studied this effect on the photosynthetic mechanism and susceptibility to photo-oxidative stress. Transgenic tobacco plants with reduced CHL P were also generated earlier by Tanaka et al. (1999). Grasses et al., (2001) in this study, have examined the significance of tocopherol content in the protection of PS II. A further reason in this study seems to be the use of CHL P antisense plants for susceptibility to photo-oxidative stress. A reduction in CHL P activity is correlated with a reduced total chlorophyll and tocopherol content along with as accumulation of geranylgeranylated Chl (Chl GG). The data of Grasses et al., (2001) are in agreement with that of Tanaka et al. (1999). The presence of Chl GG had no influence on the activity of either photosystem. However, there was a reduction of the electron transport chain per leaf area in the transgenic plants. At saturating light intensities, there was an increased photoinactivation from which CHL P antisense plants could not recover. Reduced tocopherol content is a limiting factor for defense against photo-oxidative stress. Therefore, this study has demonstrated the importance of the use of transgenic plants with reduced levels of individual antioxidants understanding light stress.

Rissler and Pogson (2001) employed an antisense novel technique resulting in reduced levels of β-hydroxylase enzyme which catalyses hydroxylation of β-rings of β-carotenes in *Arabidopsis*. Thus, their study approaches from the angle of the biosynthesis of xanthophyll carotenoids for the understanding of the photoprotection mechanisms. The antisense β-hydroxylase transgene, when expressed, has led to a substantial reduction by 64 per cent in violaxanthin and 41 per cent in neoxanthin. Under high light stress, the pool of violaxanthin was the same as in wild type; thus, there was a greater reduction of 75 per cent in zeaxanthin. This high decrease in zeaxanthin has reduced NPQ only by 16 per cent. Overall, the study of the authors has (Rissler and Pogson, 2001) led to an understanding of the expression of an antisense β-hydroxylase transgene in *A.thaliana*. The authors have conclusively demonstrated the possibility of generating plants with 64 per cent reduction in β-carotene derived xanthophylls which still support LHC assembly. The study also showed that 25 per cent of zeaxanthin is sufficient to maintain the wild type levels of NPQ. Violaxanthin is preferentially used for LHC structure optimization rather than for de-epoxidation. The study conducted by Rissler and Pogson (2001) has also confirmed the earlier work of Pogson et al., (1996) through the studies of lut 1, lut 2 that functional plasticity in xanthophylls exists in vivo.

Alia et al., (1999) have transformed *Arabidopsis thaliana* with cod A gene from *Arthrobacter*. This gene encodes choline oxidase, which converts choline to glycine betaine. The transformed plants were more tolerant to light stress than the wild type plants. The authors conclude that betaine accumulation due to Cod A gene expression has resulted in effective protection of plants from highlight stress. The transformed plants had higher ability to tolerate photoinduced inactivation of PS II. The tolerance to stress was not due to changes in membrane lipids.

Transgenics have also been used to understand the light avoidance mechanisms for photoprotection. For instance, Jeong et al., (2002) have made a study of chloroplast movement — a photoprotective measure using transgenic tobacco plants. They have generated the transgenics which possessed only 1 to 3 large chloroplasts through introduction of NtFts Z1-2, a cDNA for plastid division. It was observed that the wild type plants had normal chloroplast movement with face position under low light and profile position under high light levels. The chloroplast rearrangements under the different light regimes was completely disturbed in the transgenic plants. Accordingly, this study very clearly established the fact that smaller and larger number of chloroplasts per cell facilitates the greater light energy utilization under low light conditions.

Snyders and Kohorn (2001) have used the antisense technique in *Arabidopsis* in order to understand the function of thylakoid protein kinases (TAK's) which are required for state transitions. A structural description of photosynthesis in balancing the two photosystems has been provided by Allen and Forsberg (2001). Snyders and Kohorn (2001) have demonstrated that antisense TAK 1expression has resulted in the loss of LHCP phosphorylation and reduced state transitions. They have also shown that TAK's have a role in the regulation of thylakoid function besides state transitions; thus leading to important conclusions.

Transgenic plants have been used in understanding the nuclear genome controlled photosynthetic gene expression (Sherameti et al. 2002). Plastoquinone (PQ) redox status, however, regulates the transcription of genes located in plastid genome coding for core proteins of PS I and PS II. Oswald et al., (2001) were interested in exploring the role of plastid redox signalling in nuclear-encoded gene expression using transgenic plants. They showed that the nuclear-encoded photosynthetic genes also respond to plastid redox signals with the plastid-derived signal overriding the sugar response.

A powerful tool to establish the function of a particular compound in preventing photo-oxidative damage requires an analysis of mutants or transgenic plants, with modified levels of a single antioxidant to prevent active oxygen formation. Earlier, mutants with reduced NPQ (npq mutants) were identified in higher plants (Niyogi et al., 1998; Niyoyi, 1999). It would be of interest to examine the role of mutant studies in the understanding of photoprotection mechanism. It is also understood that light harvesting in plants is regulated by NPQ resulting in the dissipation of excess absorbed light as heat (Horton, 1996). It is considered that thermal dissipation is a crucial photoprotective mechanism which minimizes the photo-oxidative damage. NPQ depends on transthylakoid Δ pH build up in excess light, referred to as qE.

Isolation of npq 4 mutant in *Arabidopsis* has demonstrated the involvement of PsbS subunit of PS II in NPQ. In order to understand the involvement of specific xanthophylls in NPQ, Niyogi et al., (2001) have analysed the mutants affecting xanthophyll metabolism in *Arabidopsis*.

Construction of a double mutant, npq1 lut2 was done which lacked both Z and leutein due to defects in VDE and lycopene E-cyclase genes. The npq1 lut 2 strain had normal PS II efficiency at normal PS II efficiency and wild type concentrations of functional PS II RC but NPQ (q^E) was completely inhibited. The double mutant obviously had a high ability to tolerate excess light (Niyogi et al., 2001).

The results of *Arabidopsis* mutants are qualitatively similar to results obtained with equivalent mutants of *Chlamydomonas*. Structural and photophysical similarity has been noticed between Z and lutein. It is, therefore, possible that some lutein molecules bound to PS II protein perhaps PsbS may play a direct role in the ΔpH and xanthophyll dependant NPQ. The roles of xanthophylls in photoprotection has been investigated by the use of mutants of *Chlamydomonas* and *Arabidopsis* by Baroli and Niyogi (2000). From this study of mutants deficient in qE, it was noted that xanthophyll are essential but not sufficient for NPQ in high light. The quenching also requires the presence of PsbS, an intrinsic protein of PS II.

6.5. FUNCTIONAL GENOMICS OF PLANT PHOTOSYNTHESIS

The role of functional genomics in plant photosynthesis has been investigated in recent years. The implications and the potential applications of functional genomics in photosynthesis research have been reviewed in *Arabidopsis* by Pesaresi et al., (2001), in the green algae, *Chlamydomonas* by Dent et al., (2001) and Rochaix (2001a), and in the cyanobacterium, *Synechocystis* by Kaneko and Tabata (2001). It is obviously predicted that the study of photosynthesis should be considerably benefitted by the use of new genomic technologies. The functional genomics approach to photosynthesis is a very recent development and expected to yield information to identify the functions of genes and their levels of expression. A large-scale reverse genetic and tagging approaches can be used in order to assign functions to genes. Functional genomics approach uses several areas of research including, forward genetics, reverse genetics, microarray-based measurements of gene expression, 2-D electrophoresis of protein, mass spectrometry and bioinformatics (Somerville and Somerville, 1999; Pesaresi et al., 2001; Rochaix, 2001b). Using a CDNA microarray technique, Kimura et al., (2003) have surveyed 7000 genes of *Arabidopsis* in their response to high light intensity at 150 W m^{-2}. The study of Kimura et al., (2003) has established that 110 genes had a positive response to high light treatment. The genes involved in the scavenging enzymes of ROS and in the biosynthesis of lignins and flavonoids are activated by high light. Thus, the microarray system has an advantage in the screening of a large number of genes in order to identify those responsible for a particular function. It is believed that forward genetics has elucidated only a small number of gene-function relationships and the method, therefore, is comparatively a less efficient tool for the analysis of gene functions. On the other hand, reverse genetics is expected to contribute a greater understanding in gene function as evidenced by the work described earlier. A small-scale reverse genetics applied to photosynthetic function is evidenced by the work on PS I complex (Naver et al., 1999; Obokata et al., 1993). Till date, only the reverse genetics approach to photosynthesis is limited to the more well-defined genes. In this direction, an apt example is *Arabidopsis* PS I.

The area of transcriptomics is related to a discovery of the expression of genes during development in response to the environment. The redox-controlled mechanisms in the regulation of expression of photosynthesis genes were explained by Pfannschmidt et al. (2001a,b). The changes in the redox state are induced due to variations in the quality and quantity of light available. Such changes are evident during state transitions and adjustment of photosystem stoichiometry (Pfannschmidt et al., 2001a,b). Proteomics is an attempt to understand the expression, post-transcriptional control and post-translational modification of proteins.

Recently, Zolla et al., (2002) have investigated in detail the proteomic of light harvesting proteins in various species of higher plants. They have investigated the photosystem I light harvesting proteins (Lhc a) from four monocot and five dicot species by the use of liquid chromatography – electrospray ionization mass spectrometry. Their data agree with the evidence that in tomato, Lhca2 gene has the highest level of expression within all the Lhca1 genes. Lhc or 1 is the lowest abundant protein. Thus, these results have implications towards understanding the supramolecular organization of the antennae complex. Zolla et al., (2003) have extended their study of proteomics to light harvesting proteins of PS II by analyzing 14 different plant species. Their results, showing a high copy number of gene products and protein heterogeneity observed in PS II, suggested a strategy for obtaining degree of organization of light harvesting systems under different environmental conditions.

Peltier et al., (2002) have made an experimental proteome analysis with an added genome-wide prediction to understand the protein content of the thylakoid lumen using *Arabidopsis* chloroplast. Their study has established the identities of 81 proteins. It was concluded that the prime functions of lumenal proteome involved a role in folding and proteolysis of thylakoid proteins and also protection against oxidative stress. Schubert et al., (2002) have determined the proteome map of chloroplast lumen from *Arabidopsis thaliana*. Schubert et al., (2002) have shown that the thylakoid lumen has a specific proteome, of which they have identified

36 proteins. They have also observed that the thylakoid lumen from spinach contained a similar proteome. The fact that the whole proteome of lumen contained 80 proteins showed that the lumenal space is densely packed. Therefore, the studies of proteome would ultimately be of great importance in understanding the regulation of photosynthesis. Random insertional mutagenesis is a preferential strategy of functional genomics (Pesarasi et al. 2001). The obvious application of reverse genetics will be the isolation of mutants for the subunits of the photosynthetic apparatus. It is also known that the majority of genes relevant to photosynthesis identified so far by forward genetics relate to proteins targeted to the chloroplast (nuclear-coded genes). The chloroplast genome codes for over 90 proteins, while all other resident proteins are coded by nuclear genes. The chloroplast proteome in *Arabidopsis* is estimated to be 1900 to 2500 proteins. In *Arabidopsis*, chloroplast proteins account for 12 per cent of the total proteome of *Arabidopsis* and, therefore, these are the choice for study of photosynthesis by means of functional genomics (Pesaresi et al. 2001). Application of genetic engineering to the plastid genome in tobacco was demonstrated earlier by Svab and Maliga (1993).

Dent et al., (2001) have emphasized the use of *Chlamydomonas* as a model photosynthetic organism for studies on functional genomes, when compared to *Arabidopsis*, though the genome of latter is completely sequenced. It is also important to note that *Chlamydomonas* has the ability for nuclear and mitochondrial gene transformations. Photosynthetic studies using *Chlamydomonas* have made extensive application of this trait. Chloroplast transformation has not become practical in *Arabidopsis*. Chloroplast biogenesis has been investigated in *Chlamydomonas*. The subunits of proteins of photosynthetic complexes are encoded by both nuclear and chloroplast genomes. It is understood that the induction of nuclear gene expression is controlled at the transcriptional level. The chloroplast gene expression is regulated by post-transcriptional events under the control of nuclear-encoded factors (Barkan and Goldschmidt – Clermont, 2000). An example of a nuclear gene-controlling chloroplast gene expression is NAC2 gene. It is involved in the control of half life of chloroplast psbD mRNA encoding D2 of PS II RC. MBB1 gene, which encodes a NAC2 homoloque, has been isolated and is required for psbB m RNA formation. Other examples of structure, function analysis relates to photosystem I in *Chlamydomonas*. Insertional mutagenesis was used to generate a mutant in the nuclear psa F gene (Farah et al. 1995). This work proved that psa F is essential for docking of plastocyanin in PS I. Site-directed mutagenesis of psa F has shown that a single aminoacid (K 23) might represent a specific recognition site of interaction of PS I with plastocyanin. Mutants of *Chlamydomonas* defective in NPQ were isolated and the lack of xanthophyll in npq 1 mutants was the first genetic evidence for importance of Z synthesis in NPQ. The role for plastoquinol–binding site of the *Chlamydomonas* cytochrome b6f complex was demonstrated by site-directed mutagenesis. Therefore, the ultimate goal of functional genomics in photosynthesis is to understand the function of genes responsible for the synthesis, assembly, function and regulation of photosynthetic apparatus (Dent et al. 2001).

Kaneko and Tabata, (2001) have reviewed the status of functional genomics of *Synechocystis* PCC 6803, a cynobacterial system for which the entire genome has been sequenced. It is said that the genome imformation can be obtained from the data base, called cyanobase. *Synechocystis* is a unicellular cyanobacterium, used as a model organism for the study of photosynthesis. About 138 genes have been assigned roles for photosynthesis and respiration. Also, proteome and transcriptome analysis has been done to some extent in this organism and further work is expected to lead to identification of crucial genes in photosynthesis (Kaneko and Tabata, 2001). Van Wijk (2000) has made a comprehensive review of the chloroplast proteomics and has given its potential impact in providing a better understanding of the biogenesis of chloroplast, its function and adaptation.

The main purpose of this chapter has, therefore, been to highlight the use of transgenic and biotechnological approaches in the areas of research, including the subunit structural organization of photosystems, photoinhibition and photoprotection, biosynthesis of carotenoids and in chloroplast rearrangement as a photoprotective measure. The potential for the use of functional genomics and proteomics in chloroplast structure and biogenesis and in the understanding of photosynthetic performance in varying environments has been emphasized in detail.

7
Concluding Remarks

It is apparent from the foregone account that a major problem attracting the utmost attention relates to the plant stress caused by excessive light energy absorbed. The inability of plants to match the amount of light energy absorbed with that utilized in photosynthesis results in excess excitation energy. Photosynthetic capacity is the principal determinant for the utilization of harvested light energy. Light energy unutilized during photosynthesis or otherwise not dissipated through the existing photoprotective strategies produces photoinactivation of photosystem II and photoinhibition of photosynthesis (Kato et al., 2003). Therefore, a higher photosynthetic capacity or an efficient operation of different photoprotective safety valves (Niyogi, 2000) should lower the risk of photoinhibition.

Multiple strategies of photoprotection to mitigate the damaging effects of photoprotection from excess light have been identified in higher plants. These methods include the xanthophyll cycle dependent heat dissipation (NPQ), photorespiration, water-water cycle, cyclic electron transport around PS I and PS II, apart from the enhanced rates of photosynthesis and a faster recovery from photoinhibition. However, the relative contribution of each of these factors in the minimization of photoinhibition is still an enigma and requires considerable attention in the future. The role of oxygen as an alternate electron acceptor (Ort and Baker, 2002) is still debatable in view of the measured low rates of the Mehler reaction. The Cyclic electron flow around PS II (Miyake and Okamura, 2003) has been shown to reduce photoinhibition but much further work is needed in this direction so as to assess its role as a photoprotective measure.

Thermal dissipation of light energy through non-photochemical mechanism involves a major chunk of absorbed energy. It is estimated that even under adequate moisture levels and saturating levels of irradiance, thermal dissipation accounts for 60 per cent of the absorbed light, while leaf photosynthesis involves 38 per cent (see review by Flexas and Medrano, 2002). A cherished goal of research in this aspect would be to modify this partitioning of light energy and increase the proportion of the use in photochemistry of photosynthesis relative to dissipation by non-photochemical mechanisms.

While much is known regarding the occurrence, mechanism and related aspects of NPQ (Govindjee and Seufferheld, 2002), several uncertainities still exist in the knowledge of this process. A requirement of the xanthophyll cycle interconversion for the operation of NPQ is yet to be established. The multiple roles of specific xanthophylls in the assembly of LHC and in photoprotection require more intensive investigation. Also, the role of recently-proposed (Garcia – Plazoala et al., 2003) lutein epoxide cycle — besides the known xanthophyll cycle in photoprotection — needs to be examined further. Though the essentiality of PsbS subunit of PS II has been demonstrated (Li et al., 2000) in *Arabidopsis*, the involvement of other antennae proteins in the algal system have been recently implicated (Elrad et al 2002). It is also doubtful whether the PsbS subunit is pigment binding or otherwise (Dominici et al., 2003). Further work is needed in this respect to identify the precise subunits of photosystem II involved in eliciting NPQ.

Leaf and the organelle level mechanisms in the regulation of light interception require a deeper understanding. Paraheliotropism is easily understood as an adaption for mitigating water stress in view of its effects on reduced light absorption and lowered level of photoinhibition. Plants need to compromise between avoidance of water stress and the maximization of photosynthesis. However, the example of diaheliotropism in several plants (Sailaja and Das, 1996; Koller, 2000) shows that maximization of light absorption through solar tracking and improved diurnal photosynthetic rates is possible at least in these plants without photoinhiition. The manner in which the diaheliotropic plants escape the photodamage and tolerate high light levels is a subject for further work. Since these plants possess higher photosynthetic capacity, it is worth examining the uniqueness, if any, in the construction of photosynthetic apparatus in diaheliotropic plants making them suitable for high light absorption. Such a study might provide clues to understanding the utilization of higher levels of incident light energy, leading to improved photosynthesis without photodamage and further resulting in higher plant productivity.

While short-term photoprotection to excess light levels is brought about by dissipation of excess energy in the antennae and increased metabolism of reactive oxygen species, the long-term adjustment of plants to a dynamic light environment is much more complex. Considerable further work is needed to understand the underlying mechanisms of acclimation of photosynthesis to a given light regime. The precise changes of gene function and alteration in structural attributes of chloroplast suitable for photosynthesis and growth in high or low irradiance levels are needed. The mechanism of photosynthetic light acclimation in mature leaves (Sailaja and Das, 1995; Yin and Johnson, 2000; Oguchi, 2003) needs further study to understand the signifance of the plasticity of acclimation in relation to leaf development.

Gene expression studies (Teramoto et al., 2002) are needed to be carried out in higher plant systems in order to understand the regulation of the levels of LHC II proteins in response to differing light intensities. Such studies would reveal the mechanism of adjustment of antennae systems and the regulation of light harvesting in varying light levels. The gene regulation of antioxidant biosynthesis and its implications for an efficient scavenging of active oxygen is another area of fruitful research in photoprotection of photosynthesis.

References

Abdallah F, Salamini F, Leister D, 2000, A prediction of the size and evolutionary origin of the proteome of chloroplasts of *Arabidopsis*. Trends Plant Sci. **5**: 141-142.

Adam Z, Adamska I, Nakabayashi K, Ostersetzer O, Haussuhl K, Manuell A, Zheng B, Vallon O, Rodermel SR, Shinozaki K, Clarke AK, 2001, Chloroplast and mitochondrial proteases in *Arabidopsis thaliana* : a proposed nomenclature. Plant Physiol. **125**: 1912-1918.

Adir N, Showchat S, Ohad I, 1990, Light dependent D1 protein synthesis and translocation is regulated by reaction center II : reaction center II serves as an acceptor for the D1 precursor. J. Biol. Chem. **265**: 12563-12568.

Agati G, Mazzinghi P, Fusi F, Ambrosini I, 1995, The F 685/F 730 chlorophyll fluorescence ratio as a tool in plant physiology: response to physiological and environmental factors. J. Plant Physiol. **145**: 228-238.

Albertsson PA, 2001, A quantitative model of the domain structure of the photosynthetic membrane. Trends Plant Sci. **6**: 349-353.

Alia, Kondo Y, Sakamoto A, Nonaka M, Hayashi H, Pardha Saradhi P, Chen THT, Murata N, 1999, Enhanced tolerance to light stress of transgenic *Arabidopsis* plants that express the cod A gene for a bacterial choline oxidase. Plant Mol. Biol. **40**: 279-288.

Allen JF, Forsberg J, 2001, Molecular recognition in thylakoid structure and function. Trends Plant Sci. **6**: 317-326.

Alves PL da CA, Magalhaes ACN, Barja PR, 2002, The phenomenon of photoinhibition of photosynthesis and its importance in reforestation. Bot. Rev. **68**(2): 193-208.

Ananyev GM, Sakiyan I, Diner BA, Dismukes GC, 2002, A Functional role for Tyrosine–D in assembly of the inorganic core of the water oxidase complex of photosystem II and the kinetics of water oxidation. Biochemistry **41**: 974-980.

Anderson JM, 1986, Photoregulation of the composition, function, and structure of thylakoid membranes. Annu. Rev. Plant Physiol. **37**: 93-136.

Anderson JM, Aro EM, 1994, Grana stacking and protection of photosystem

II in thylakoid membranes of higher plant leaves under sustained high irradiances : An. Hypothesis Photosynth. Res. **41**: 315-326.

Anderson JM, Chow WS, Park Y-II, 1995, The grand design of photosynthesis: acclimation of the photosynthetic apparatus to environmental cues. Photosynth. Res. **46**: 129-139.

Anderson JM, Chow WS, Park Y-II, Franklin LA, Robinson SPA, Van Hasselt PR, 2001, Response of *Tradesuantia albiflora* to growth irradiance: change versus changeability. Photosynth. Res. **67**: 103-112.

Anderson JM, Osmond CB, 1987, Shade-sun responses: compromises between acclimation and photoinhibition. In: DJ Kyele, CB Osmond, CJ Arntzen, (eds) Photoinhibition: Elsevier, Amsterdam pp.1-38.

Anderson JM, Park Y-II, Chow WS, 1997a, Photoinactivation and photoprotection of photosystem II in nature. Physiol. Plant **100**: 214-223.

Anderson JM, Park Y-II, Chow WS, 1998, Unifying model for the photoinactivation of photosystem II *in vivo* under steady state photosynthesis. Photosynth. Res. **56**: 1-13.

Anderson JM, Price GD, Chow WS, Hope AB, Badger MR, 1997b, Reduced levels of cytochrome bf complex in transgenic tobacco leads to marked photochemical reduction of the plastoquinone pool, without significant change in acclimation to irradiance. Photosynth. Res. **53**: 215-227.

Andersson B, Aro EM, 2001, Photodamage and D1 protein turnover in photosystem II. In: EM, Aro, B Andersson, eds, Advances in Photosynthesis and Respiration, Vol. 11. Kluwer Academic Publishers, Dordrecht, pp.377-393.

Andersson B, Barber J, 1996, Mechanisms of photodamage and protein degradation during photoinhibition of photosystem II. In: NR Baker, (ed) Photosynthesis and the Environment. Kluwer Academic Publishers, Netherlands, pp.101-121.

Andersson B, Ponticos M, Barber J, Koivuniemi A, Aro EM, Hagman A, Salter AH, Dan Hui Y, Lindahl M, 1994, Light induced proteolysis of photosystem II reaction center proteins and LHC II in isolated preparations. In: NR Baker, JR Bowyer, (eds) Photoinhibition of Photosynthesis: From Molecular Mechanisms to the Field, Bios Scientific Publishers, Oxford pp.143-159.

Andersson J, Walters RG, Horton D, Jansson S, 2001, Antisense inhibition of the Photosynthetic antennae proteins CP29 and CP26: implications for the mechanism of protective energy dissipation. Plant Cell, **13**: 1193-1204.

Apel K, 2001, Chlorophyll Biosynthesis-Metabolism and strategies of higher plants to avoid photo-oxidative stress. In: EM Aro, B Anderson, (eds) Advances in Photosynthesis and Respiration. Vol.11, Kluwer Academic Publishers, Dordrecht. pp.235-252

Aro EM, McCafferi S, Anderson JM, 1993a, Photoinhibition and D1 protein degradation in peas acclimated to different growth irradiances. Plant Physiol. **103**: 835-843.

Aro EM, Virgin I, Andersson B, 1993b, Photoinhibition of photosystem II. Inactivation, protein damage and turnover Biochem. Biophys. Acta. **1143**: 113-134.

Asada K, 1992, Ascorbate peroxidase – A hydrogen peroxide – Seavenging enzyme in plants. Physiol Plant. **85**: 235-241.

Asada K, 1994, Mechanisms for scavenging reactive molecules generated in chloroplasts under light stress. In: NR Baker, JR Bowyer, (eds) Photoinhibition of Photosynthesis. Bios. Sci. Publishers, Oxford, pp.129-141.

Asada K, 1996, Radical production and scavenging in the chloroplasts In: NR Baker, ed, Photosynthesis and the Environment Kluwer Academic publishers, pp.123-150.

Asada K, 1997, The role of ascorbate peroxidase and monodehydro ascorbate reductase in H_2O_2 seavenging in plants. In: JG Scandilos, (ed), oxidative stress and Molecular Biology of Antioxidant Defenses. Cold Spring Harbor Lab. Press New York pp.527-568.

Asada K. 1999, The water-water cycle in chloroplasts: Scavenging of active oxygens and dissipation of excess photons. Annu. Rev. Plant Physiol. Plant Mol. Biol, **50**: 601-637.

Asada K, Kiso K, Yoshikawa K, 1974, Univalent reduction of molecular oxygen by spinach chloroplasts on illumination. J. Biol. Chem. **249**: 2175-2181.

Augusti A, Scartazza A, Navari-Izzo F, Sgherri CLM, Stevanovic B, Brugnoli E, 2001, Photosystem II photochemical eficiency, zeaxanthin and antioxidant contents in the poikilohydric *Ramonda serbica* during dehydration and rehydration. Photosynth. Res. **67**: 79-88.

Badger M, Von Caemmerer S, Ruuska S, Nakano H, 2000, Electron flow to oxygen in higher plants and algae: rates and control of direct photoreduction (Mehler reaction) and rubisco oxygenase. Phil. Trans. R. Soc. Lond. B. **355**: 1433-1446.

Baena–Gonzalez E, Barbato R, Aro EM, 1999, Role of phosphorylation in the repair cycle and oligomeric structure of photosystem II. Planta. **208**: 196-204.

Bailey S, Thompson E, Nixon PJ, Morton P, Mullineaux CW, Robinson C, Mann NH, 2002, A critical role for the var 2 FtsH homologue of *Arabidopsis thaliana* in the photosystem II repair cycle *in* vivo. J. Biol. Chem. (277) **3**: 2006-2011.

Bailey S, Walters RG, Jansson S, Horton P, 2001, Acclimation of *Arabidopsis thaliana* to the light environment: The existence of separate low light and high light responses. Planta **213**: 794-801.

Baker NR, 1996, Environmental constraints on photosynthesis. An overview of some future prospects. In: NR Baker, (ed) Photosynthesis and the Environment, Kluwer Academic Publishers, Dordrecht, pp.469-476.

Barbato R, Friso G, Polvasino de Laureto P, Frizzo A, Rigoni F, Giacometti GM, 1992a, Light-induced degradation of D2 protein in isolated photosystem II reaction center complex. FEBS Lett. **311**: 33-36.

Barbato R, Friso G, Rigoini F, Frizzo A, Giacometti GM, 1992c, Characterisation of a 41 kDa photoinhibition adduct in isolated photosystem II reaction centers. FEBS Lett. **309**: 165-169.

Barbato R, Frizzo A, Friso G, Rigoni F, Giacometti GM, (1992b), Photoinduced degradation of the D1 protein in isolated thylakoids and various photosystem II particles after donor side inactivation. FEBS Lett. **304**: 136-140.

Barber J, 2002, Photosystem II: A multisubunit membrane protein that oxidizes water. Curr. Opin. Struct. Biol. **12**: 523-530.

Barber J, Andersson B, 1992, To much of a good thing: light can be bad for photosynthesis. Trends Biochem. Sci. **17**: 61-66.

Barber J, Kuhlbrandt W, (1999), Photosystem II. Curr. Opin. Struct. Biol. **9**: 469-475.

Barber J, Nield J, Morris EP, Hankamer B, (1999) Subunit positioning in photosystem II revisited. Trends Biochem. Sci. **24**: 43-45.

Barber J, Nield J, Morris EP, Zheleva D, Hankamer B, 1997, The structure function and dynamics of photosystem II. Physiol. Plant **100**: 817-827.

Barkan A, Goldschmidt Clermont M, 2000, Participation of nuclear genes in chloroplast gene expression. Biochimie. **82**: 559-572.

Baroli I, Niyogi KK, 2000, Molecular genetics of xanthophyll-dependent photoprotection in green algae and plants. Phil. Trans. R. Soc. Lond. B. **355**: 1385-1394.

Barry BA, Boerner RJ, de Paula JC, 1994, The use of cyanobacteria in the study of the structure and function of photosystem II. In: DA Bryant, (ed) The Molecular Biology of Cyanobacteria. Kluwer Academic Publishers pp.217-57.

Barter LMC, Bianchietti M, Jeans C, Schilstra MJ, Hankamer B, Diner BA, Barber J, Durrant JR, Klug DR, 2001, Relationship between excitation energy transfer, trapping, and antennae size in photosystem II. Biochemistry **40**: 4026-4034.

Barth C, Krause GH, 2002, Study of tobacco transformants to assess the role of chloroplastic NAD(P)H dehydrogenase in photoprotection of photosystem I and II. Planta. **216**: 273-279.

Bassi R, Dianese P, 1992, A supramolecular light harvesting complex from chloroplast photosystem II membranes. Eur. J. Biochem. **204**: 317-26.

Bassi R, Pineau B, Dianese P, Marquardt J, 1993, Carotenoid binding proteins of photosystem II. Eur. J. Biochem. **212**: 297-303.

Bassi R, Rigoni F, Giacometti GM, 1990, Chlorophyll binding proteins with antennae function in higher plants and green algae. Photochem. Photobiol. **52**: 487-1206.

Bassi R, Sandona D, Croce R, 1997, Novel aspects of chlorophyll a/b binding proteins. Physiol. Plant **100**: 769-779.

Batschauer A, 1998, Photoreceptors of higher plants. Planta **206**: 479-492.

Bell CJ, Rose DA, 1981, Light measurement and terminology of flow. Plant Cell Environ. **4**: 89-96.

Bennoun P, 1982, Evidence for a respiratory chain in the chloroplast. Proc. Natl Acad. Sci. USA **79**: 4352-56.

Biehler K, Fock H, 1996, Evidence for the contribution of the Mehler – Peroxidase reaction in dissipating excess electrons in drought-stressed wheat. Plant Physiol. **112**: 265-272.

Bilger W, Bjorkman O, 1994, Relationships among violaxanthin de-epoxidation thylakoid membrane conformation and nonphotochemical chlorophyll fluorescence quenching in leaves of cotton *Gossypium hirsutum* L. Planta. **193**: 238-46.

Bjorkman O, Demmig-Adams B, 1995, Regulation of photosynthetic light energy capture, conversion and dissipation in leaves of higher plants. In: ED Schulze, MM Caldwell, (eds) Ecophysiology of Photosynthesis. Springer-verlag, Berlin, pp.16-45.

Blubagh DJ, Atamian M, Babcock GT, Goldbeck JH, Cheniae GN, 1991, Photoinhibition of hydroxylamine–extracted photosystem II membranes : Identification of sites of damage. Biochemistry **30**: 7586-7597.

Boardman NK, 1977, Comparative photosynthesis of sun and shade plants. Annu. Rev. Plant. Physiol. **28**: 355-377.

Boekema EJ, Hankamer B, Bald D, Kruip J, Nield J, Boonstra AF, Barber J, Rogner M, 1995, Supramolecular structure of the photosystem II complex from green plants and cyanobacteria Proc. Natl. Acad. Sci. USA **92**: 175-179.

Bratt CE, Arvidsson PO, Carlsson M, Akerlund HE, 1995, Regulation of violaxanthin de-epoxidase activity by pH and ascorbate concentration. Photosynth. Res. **45**: 169-175.

Brettel K, 1997, Electron transfer and redox cofactors in photosystem I. Biochim. Biophys. Acta. **1318**: 322-373.

Bricker TM, Frankel LK, 2002, The structure and function of CP47 and CP43 in photosystem II. Photosynth. Res. **72**: 131-146.

Bricker TM, Ghanotakis DF, (1996), Introduction to oxygen evolution and the oxygen evolving complex. In: DR Ort, CF Yocum, (eds) Advances in Photosynthesis Vol.4 Kluwer Academic Publishers, Dordrecht, pp.113-136.

Briggs WR, Christie JM, 2002, Phototropins 1 and 2: Versatile plant blue – light receptors. Trends Plant Sci. **1**: 204-210.

Britton G, (1995), Structure and properties of carotenoids in relation to function. FASE B. J. **9**: 1551-1558.

Brugnoli E, Bjorkman O, 1992, Chloroplast movement in leaves : Influence on chlorophyll fluorescence and measurement of light – induced absorbance changes related to Δ pH and zeaxanthin formation. Photosynth. Res. **32**: 23-35.

Buch K, Stransky H, Hager A, 1995, FAD is a further essential cofactor of the NAD(P)H and O_2- dependent zeaxanthin – epoxidase. FEBS Lett. **376**: 45-48.

Bugos RC, Chang SH, Yamamoto HY, 1999, Developmental expression of violaxanthin de-epoxidase in leaves of tobacco growing under high and low light. Plant Physiol. **124**: 207-213.

Bugos RC, Hieber AD, Yamamoto HY, 1998, Xanthophyll cycle enzymes are members of the lipocalin family, the first identified from plants. J. Biol. Chem. **273**: 15321-15324.

Bukhov NG, Heber U, Wiese C, Shuvalov VA, 2001a, Energy dissipation in photosynthesis: Does the quenching of chlorophyll fluorescence originate from antennae complexes of photosystem II or from the reaction centre? Planta. **212**: 749-758.

Bukhov NG, Kopecky J, Pfündel EE, Klughammer C, Heber U, 2001b, A few molecules of zeaxanthin per reaction centre of photosystem II permit effective thermal dissipation of light energy in photosystem II of a poikilohydric moss. Planta. **212**: 739-748.

Burrows PA, Sazanov LA, Svab Z, Maliga P, Nixon PJ, 1998, Identification of a functional respiratory complex in chloroplasts through analysis of tobacco mutants containing disrupted plastid ndh genes. EMBO J. **17**: 868-876.

Caffarri S, Croce R, Breton J, Bassi R, 2001, The major antennae complex of photosystem II has a xanthophyll binding site not involved in light harvesting. J. Biol. Chem. **276**: 35924-35933.

Cai SQ, Chen GY, Zhang HB, Xu DQ, 2002, Saturating irradiance induced photoinhibition without monomerisation of photosystem 2 dimers in soyabean leaves. Photosynthetica. **40**: 215-220.

Canovas PM, Barber J, 1993, Detection of a 10 kDa breakdown product containing the C-terminus of the D1 protein in photoinhibited wheat leaves suggests an acceptor side mechanism. FEBS Lett. **324**: 341-344.

Carol P, Stevenson D, Bisanz C, Breitenbach J, Sandaman G, 1999, Mutations in the *Arabidopsis* gene immutans cause a variegated phenotype by inactivating a chloroplast terminal oxidase associated with phytoene desaturation. Plant Cell **11**: 57-68.

Casano LM, Martin M, Sabater B, 2001, Hydrogen peroxide mediates the induction of chloroplastic Ndh complex under photooxidative stress in Barley. Plant Physiol. **125**: 1450-1458.

Casano LM, Martin M, Zapata JM, Sabater B, 1999, Leaf age and paraquat concentration dependent effects on the levels of enzymes protecting against photooxidative stress. Plant Sci. **149**: 13-22.

Chappelle EW, Mc Mutry JE, Woud FM, Newcomb WW, 1984, Laser induced fluorescence of green plants 2: LIF caused by nutrient deficiencies in corn. Applied Optics **23**: 139-142.

Chassin Y, Kapri-Parde E, Sinvany G, Arad T, Adam Z, 2002, Expression and characterization of the thylakoid lumen protease Deg P1 from *Arabidopsis.* Plant Physiol. **130**: 857-864.

Chitnis PR, 2001, Photosystem I: Function and physiology. Annu. Rev. Plant, Physiol. Plant Mol. Biol. **52**: 593-626.

Chow WS, Anderson JM, 1987a, Photosynthetic responses of *Pisum sativum* to an increase in irradiance during growth I. Photosynthetic activities. Aust. J. Plant Physiol. **14**: 1-8.

Chow WS, Anderson JM, 1987b. Photosynthetic responses of *Pisum sativum* to an increase in irradiance during growth II. Thylakoid membrane components. Aust. J. Plant Physiol. **14**: 9-19.

Chow WS, Funk C, Hope AB, Govindjee, 2000, Greening of intermittent light grown bean plants in continuous light: Thylakoid components in relation to photosynthetic performance and capacity for photoprotection. Indian J. Biochem. Biophys. **37**: 395-404.

Chow WS, Melis A, Anderson JM, 1990, Adjustments of photosystem stoichiometry in chloroplasts improve the quantum efficiency of photosynthesis. Proc. Natl. Acad. Sci, USA **87**: 7502-7506.

Christie JM, Briggs WR, 2001, Blue light sensing in higher plants. J. Biol. Chem. **276**(15): 11457-11460.

Clarke JE, Johnson GN, 2001, *In vivo* temperature dependence of cyclic and pseudocyclic electron transport in barley. Planta **212**: 808-816.

Cleland RC, Melis A, 1987, Probing the events of photoinhibition by altering electron-transport activity and light-harvesting capacity in chloroplast thylakoids. Plant Cell Environ. **10**: 747-752.

Cornic G, Bukhov NG, Wiese C, Bligny R, Heber U, 2000, Flexible coupling between light-dependent electron and vectorial proton transport in illuminated leaves of C3 plants. Role of photosystem I dependent proton pumping. Planta. **210**: 468-477.

Creighton AM, Hulford A, Mant A, Robinson D, Robinson C, 1995, A monomeric tightly folded stromal intermediate on the delta pH – dependent thylakoidal protein transport pathway. J. Biol. Chem. **270**: 1663-1669.

Creissen G, Firmin J, Fryer M, Kular B, Leyland N, Reynolds H, Pastori G, Wellburn F, Baker N, Wellburn A, Mullineaux, 1999, Elevated glutathione biosynthetic capacity in the chloroplasts of transgenic tobacco plants paradoxically causes increased oxidative stress. Plant Cell. **11**: 1277-1291.

Critchley C, 1998, Photoinhibition. In: AS, Raghavendra (ed) Photosynthesis. Cambridge University Press pp.264-272.

Croce R, Weiss S, Bassi R, 1999, Carotenoid binding sites of the major light-harvesting complex II of higher plants. J. Biol. Chem. **274**: 29613-29623.

Crofts AR, Yerkes CT, 1994, A molecular mechanism for qE quenching FEBS Lett. **352**: 265-270.

Cronlund SL, Forseth IN, 1995, Heliotropic leaf movement response to H^+ATPase activation, H^+ATPase inhibition and K^+ channel inhibition *in vivo*. Am. J. Bot. **82**: 1507-1513.

Dalbey RE, Robinson C, 1999, Protein translocation into and across the bacterial plasma membrane and the plant thylakoid membrane. Trends Biochem. Sci. **24**: 17-22.

Danon A, Mayfield SP, 1994, ADP – dependent phosphorylation regulates RNA – binding *in vitro*. Implication in light mediated translation. EMBO J. **13**: 2227-2235.

Davies EC, Chow WS, Le Fay JM, Jordan BR, (1986), Acclimation of tomato leaves to changes in light intensity. Effects on the function of the thylakoid membrane. J. Exp. Bot. **37**: 211-220.

De la Torre WR, Burkey KO, 1990, Acclimation of barley to changes in light intensity: Photosynthetic electron transport activity and components. Photosynth. Res. **24**: 127-136.

De Las Rivas J, Andersson B, Barber J. 1992, Two sites of primary degradation of the D1 protein induced by acceptor or donor side photoinhibition in photosystem II core complexes. FEBS Lett **30**: 246-252.

De Las Rivas J, Shipton CA, Ponticos M, Barboc J, 1993, Acceptor side mechanism of photoinduced proteolysis of the D1 protein in photosystem II reaction centers. Biochemistry **32**: 6944-6950.

Debus RJ, 2000, The polypeptides of photosystem II and their influence on manganotyrosyl based oxygen evolution. In: A Siegel, H Siegel, (eds), Metal Ions in Biological Systems, Marcel Dekker, New York pp. 657-711.

DeCoster B, Christensen RL, Gebhard R, Lugtenburg J, Farhoosh R, Frank HA, 1992, Low lying electronic states of carotenoids. Biochim. Biophys. Acta. **1102**: 107-109.

Demmig–Adams B, 1990, Carotenoids and photoprotection in plants: a role for the xanthophyll – zeaxanthin. Biochim. Biophys. Acta. **1020**: 1-24.

Demmig-Adams B, Adams III WW, 1992, Photoprotection and other responses of plants to high light stress. Annu. Rev. Plant Physiol. Plant. Mol. Biol, **43**: 599-626.

Demmig-Adams B, Adams III WW, 1996a, Chlorophyll and carotenoid composition in leaves of *Euonymus kiautschovicus* acclimated to different degrees of light stress in the field. Aust. J. Plant Physiol. **23**: 649-659.

Demmig-Adams B, Adams III WW, 1996b, The role of xanthophyll cycle carotenoids in the protection of photosynthesis. Trends Plant Sci. **1**: 21-26.

Demmig-Adams B, Adams III WW, 2000, Harvesting sunlight safely. Nature **403**: 371-374.

Demmig-Adams B, Adams III WW, Winter K, Meyer A, Sehreiber U, Pereira JS, Kruger A, Crygan F-C, Lang OL, 1989, Photochemical efficiency of photosystem II, photon yield of O_2 evolution, photosynthetic capacity and carotenoid composition during midday depression of net CO_2 uptake in *Arbutus unedo* growing in portugal. Planta **177**: 377-380.

Demmig-Adams B, Adams III WW, Baker DH, Logan BA, Bowling DR, Verhoeven AS, 1996, Using chlorophyll fluorescence to assess the fraction of absorbed light allocated to thermal dissipation of excess excitation. Physiol. Plant **98** : 253-264.

Demmig-Adams B, Adams III WW, Ebbert V, Logan BA, 1999, Ecophysiology of the xanthophyll cycle. In: HA Frank, AJ Young, G

Britton, RJ Cogdell, (eds) Advances in Photosynthesis. Vol.8 Kluwer Academic Publishers, Dordrecht, pp.245-269.

Demming–Adams B, Adams III WW, 2002, Antioxidants in photosynthesis and human nutrition. Science **298**: 2149-2153.

Dent RM, Han M, Niyogi KK, 2001, Functional genomics of plant photosynthesis in the fast lane using *Chlamydomonas reinhardtii*. Trends Plant Sci. **6**: 364-371.

Depka B, Jahns P, Trebst A, 1998, β-carotene to zeaxanthin conversion in the rapid turnover of the D1 protein of photosystem II. FEBS Lett. **424**: 267-270.

Dietz KJ, Horling F, Konig J, Baier M, 2002, The function of the chloroplast 2-cysteine peroxiredoxin in peroxide detoxification and its regulation. J. Exp. Bot. **53**: 1321-1329.

Diner BA, Rappaport F, 2002, Structure, dynamics and energetics of the primary photochemistry of phosystem II of oxygenic photosynthesis. Annu. Rev. Plant Biol. **53**: 551-580.

Diner BA, Schlodder E, Nixon PJ, Coleman WJ, Rappaport F, Lavergne J, Vermaas WFJ, Chisholm DA, 2001, Site-directed mutations at D1-His 198 and D2-His 197 of photosystem II in *Synechocystis* PCC 6803. Sites of primary charge separation and cation and triplet stablization. Biochemistry **40**: 9265-9281.

Dominici P, Caffarri S, Armenante F, Ceoldo S, Crimi M, Bassi R, 2002, Biochemical properties of the PsbS subunit of photosystem II either purified from chloroplast or recombinant. J. Biol. Chem. **277**: 22750-22758

Donahue R, Berg VS, 1990, Leaf orientation of soybean seedlings II: Receptor sites and light stimuli. Crop Science **30**: 638-643.

Ebbert V, Demmig-Adams B, Adams III WW, Mueh KE, Staehelin LA, 2001, Correlation between persistent forms of zeaxanthein-dependent energy dissipation and thylakoid protein phosphorylation. Photosynth. Res. **67**: 63-78.

Eckardt NA, 2001, A role for PsbZ in the core complex of photosystem II Plant Cell **13**: 1245-1248.

Eckert HJ, Geiken B, Bernarding J, Napiwotski A, Eichler HJ, Ringer G, 1991, Two sites of photoinhibition of the electron transfer in oxygen evolving and tris – treated PS II membrane fragments from spinach. Photosynth. Res. **27**: 97-103.

Ehleringer J, Forseth I, 1980, Solar tracking by plants. Science **210**: 1094-1098.

Elrad D, Niyogi KK, Grossman AR, 2002, A major light-harvesting polypeptide of photosystem II functions in thermal dissipation. Plant Cell **14**: 1801-1816.

Endo T, Shikanai T, Takabayashi A, Asada K, Sato F, 1999, The role of chloroplastic NAD(P)H dehydrogenase in photoprotection. FEBS Lett. **457** : 5-8.

Escoubas JM, Lomas M, La Roche J, Falkowshi PG, 1995, Light intensity regulation of *cab* gene transcription is signaled by the redox state of the plastoquinol pool. Proc. Natl. Acad. Sci. USA **92**: 10237-10241.

Eshagi S, Andersson B, Barber J, 1999, Isolation of a highly active PS II – LHC II supercomplex from thylakoid membranes by a direct method. FEBS Lett. **446**: 23-26.

Eskling M, Akerlund HE, 1998, Changes in the quantities of violaxanthin de-epoxidase, xanthophylls and ascorbate in spinach upon shift from low to high light. Photosynth. Res. **57**: 41-50.

Eskling M, Andersson PO, Akerlund HE, 1997, The xanthophyll cycle, its regulation and components. Physiol. Plant. **100**: 806-816.

Eskling M, Emanuelsson A, Akerlund HE, 2001, Enzymes and mechanisms for violaxanthin-zeaxanthin conversion. In: EM Aro, B Andersson, (eds) Advances in Photosynthesis and Respiration, Vol.11, Kluwer Academic Publishers, Dordrecht, pp.433-452.

Evans JR, 1996, Developmental constraints on photosynthesis: Effects of light and nutrition. In: NR Baker, (ed.) Photosynthesis and the Environment. Kluwer Academic Publishers, The Netherlands, pp.281-304.

Falkowski PG, Laroche J, 1991, Acclimation to spectral irradiance. J. Phycol. **27**: 8-14.

Faller P, Debus RJ, Brettel K, Sugiura M, Rutherford AW, Bonssac A, 2001, Rapid formation of the stable tyrosyl radical in photosystem II. Proc. Natl. Acad. Sci. USA **98**: 14368-14373.

Farah J, Rappaport F, Choquet Y, Joliet P, Rochaix JD, 1995, Isolation of a PsaF–deficient mutant of *Chlamydomonas reinhardtii*: efficient interaction plastocyanin with the photosystem I reaction center is mediated by the PsaF subunit. EMBO J. **14**: 4976-4984.

Farber A, Young AJ, Ruban AV, Hotron P, Jahns P, 1997, Dynamics of xanthophyll – cycle activity in different antenna sub complexes in the photosynthetic membranes of higher plants. Plant Physiol. **115**: 1609-161D.

Ferrar PJ, Osmond CB, 1986, Nitrogen supply as a factor influencing photoinhibition and photosynthetic acclimation after transfer of shade grown *Solanum dulcamara* to bright light. Planta. **168**: 563-570.

Field TS, Nedbal L, Ort DR, 1998, Non-photochemical reduction of the plastoquinone pool in sunflower leaves originates from chlororespiration. Plant Physiol. **116**: 1209-1218.

Finazzi G, Barbagallo RP, Bergo E, Barbato R, Forti G, 2001, Photoinhibition of *Chlamydomonas reinhardtii* in state 1 and state 2. J. Biol. Chem. **276**: 2251-2257.

Fischer M, Funk E, Steinmiiller K, 1997, The expression of subunits of the mitochondrial complex I–homologous NAD(P)H-plastoquinone–oxidoreductase during plastid development. Z. Naturforschung. C. **52**: 481-486.

Flachman R, 1997, Composition of Photosystem II antennae in light harvesting complex II antisense tobacco plants at varying irradiances. Plant Physiol. **113**: 787-794.

Flexas J and Medrano H, 2002, Energy dissipation in C3 plants under drought. Funct. Plant Biol. **29**: 1209-1215.

Flexas J, Hendrickson L, Chow WS, 2001, Photoinactivation of photosystem II in high light-acclimated grapevines. Aust. J. Plant Physiol. **28**: 755-764.

Formaggio E, Cinque G, Bassi R, 2001, Functional architecture of the major light harvesting complex from higher plants. J. Mol. Biol. **314**: 1157-1166.

Forseth IN, 1990, Function of leaf movements. In: RL Satter, HL, Gorton TC Vogelmann, eds, The Pulvinus: Organ for leaf movement. American Society of Plant Physiologists. Rockville pp.238-261.

Forseth IN, Ehleringer JR, 1983, Ecophysiology of two solar tracking desert annuals IV. Effects of leaf orientation in calculated daily carbon gain and water use efficiency. Oecologia **58**: 10-18.

Forster B, Osmond CB, and Boynton JE, 2001, Very high light resistant mutants of *Chlamydomonas reinhardtii*: Responses of photosystem II, non photochemical quenching and xanthophyll pigments to light and CO_2. Photosynth. Res. **67**: 5-15.

Fotinou C, Ghanotakis DF, 1990, A preparative method for the isolation of the 43 kDa, 47 kDa and the D1- D2 – cyt b 559 species directly from thylakoid membranes. Photosynth. Res. **25**: 141-145.

Foyer C, Osmond B, Walker D, 2000, Introduction Phil. Trans. R. Soc. Lond. B. **355**: 1333-1335.

Foyer CH, Harbinson J, 1999, Relationships between antioxidant metabolism and carotenoids in the regulation of photosynthesis. In: AH Frank, AJ Young, G Britton, (eds), Advances in Photosynthesis Vol.8, Kluwer Academic Publishers, Dordrecht. pp.305-325.

Foyer CH, Lopez – Delgado H, Dat JF, Scott IM, 1997, Hydrogen peroxide and glutathione–associated mechanisms of acclimatory stress tolerance and signaling. Physiol Plant. **100** : 241-254.

Foyer CH, Noctor G, 2000, Oxygen processing in photosynthesis regulation and signaling. New Phytol. **146** : 359-388.

Frank HA, Bautista JA, Josue JS, Young AJ, 2000, Mechanism of nonphotochemical quenching in green plants : Energies of the lowest excited singlet states of violaxanthin and zeaxanthin. Biochemistry **39** : 2831-2837.

Frank HA, Cua A, Chynwat V, Young AJ, Gosztola D, Wasielewski MR, 1994, Photophysics of the carotenoids associated with the xanthophyll cycle in photosynthesis. Photosynth. Res. **41** : 389-95.

Frechilla S, Zhu J, Talbott LD, Zeiger E, 1999, Stomata from npql, a zeaxanthin–less *Arabidopsis* mutant lack a specific response to blue light. Plant Cell Physiol. **40** : 949-954.

Fryer MJ, Andrews JR, Oxborough K, Blowers DA, Baker NL, 1998, Relationship between CO_2 assimilation, photosynthetic electron transport, and active O_2 metabolism in leaves of maize in the field during periods of low temperature. Plant Physiol. **116**: 571-580.

Fryer MJ, 1992, The antioxidant effects of thylakoid vitamin E (α - tocopherol) Plant Cell Environ. **15** : 381-392.

Fugita Y, Murakami A, Oki K, 1987, Regulation of photosynthetic composition in the cyanobacterial photosynthetic system: the regulation occurs in response to the redox state of the electron pool between the photosystems. Plant Cell Physiol. **28**: 283-292.

Funk C, 2001, The PsbS protein a cab-protein with a function of its own. In: EM Aro, B.Andersson, (eds), Advances in Photosynthesis and Respiration. Vol.11, Kluwer Academic Publishers Dordrecht pp.453-467.

Funk C, Adamska I, Green BR, Anderson B, Renger G, 1995, The nuclear –encoded chlrophyll–binding photosystem II–s protein is stable in the absence of pigments. J. Biol. Chem. **270**: 30141-30147.

Ganetag U, Strand A, Gustafsson D, Jansson S, 2001, The properties of the chlorophyll a/b-binding proteins Lhca2 and Lhca3 studied in vivo using antisense inhibition. Plant Physiol. **127**: 150-158.

Garab G, Cseh Z, Kovacs L, Rajagopal S, Varkonyi Z, Wentworth M, Mustardy L, Der A, Ruabn AV, Papp E, Holzenburg A, Horton P, 2002, Light – induced trimer to monomer transition in the main light-harvesting antennae complex of plants : Thermo-optic mechanism. Biochemistry **41**: 15121-15129.

Garcia – Plazaola JI, Hernandez A, Olano JM, Becerril JM, 2003, The operation of the lutein epoxide cycle correlates with energy dissipation. Funct. Plant Biol. **30** : 319-324.

Garcia-Plazaola JI, Artetxe U, Becirill JM, 1999, Diurnal changes in antioxidant and carotenoid composition in the Mediterranean sclerophyll tree, *Quercus ilex* (L) during winter. Plant Sci. **143** : 125-133.

Genty B, Harbinson J, 1996, Regulation of light utilization for photosynthetic electron transport. In: NR Baker, (ed.) Photosynthesis and the Environment, Kluwer Academic Publishers, Amsterdam pp.69-99.

Gerst U, Schreiber U, Neimanis S, Heber U, 1995, Photosystem I-dependent cyclic electron flow contributes to the control of photosystem II in leaves when stomata close under water stress. In: P.Mathis, (ed.) Photosynthesis : From light to biosphere, Vol.II. Kluwer Academic Publishers, Netherlands, pp. 835-838.

Ghanotakis DF, Yocum CF, 1986, Purification and properties of an oxygen evolving reaction center from photosystem II membranes. FEBS Lett. **197**: 244-288.

Gilmore AM, Hazlett TL, Debrunner PG, Govindjee, 1996, Photosystem II chlorophyll a fluorescence lifetimes and intensity are independent of the antennae size differences between barely wild-type and chlorina mutants :Photochemical quenching and xanthophyll cycle – dependant nonphotochemical quenching of fluorescence. Photosynth. Res. **48** : 171-187.

Gilmore AM, Yamamoto HY, 1993, Linear models relating xanthophylls and lumen acidity to non-photochemical fluorescence quenching. Evidence that antheraxanthin explains zeaxanthin – independent quenching. Photosynth. Res. **35** : 67-78.

Gilmore AM, 1997, Mechanistic aspects of xanthophyll cycle-dependent photoprotection in higher plant chloroplasts and leaves. Physiol. Plant. **99**, 197-209.

Gilmore AM, 2001, Xanthophyll cycle dependent nonphotochemical quenching in photosystem II: Mechanistic insights gained from *Arabidopsis thaliana* L. mutants that lack violaxanthin de-epoxidase activity and/or lutein. Photosynth. Res. **67**: 89-101.

Gilmore AM, Govindjee, 1999, How higher plants respond to excess light: Energy dissipation in photosystem II In: GS Singhal, G Renger, SK Sopory, KD Irrgang, Govindjee, (eds) Narosa Publishing House, New Delhi, pp.513-547.

Gilmore AM, Shinkarev VP, Hazlett TH, Govindjee (1998) Quantitative analysis of the effects of intrathylakoid pH and xanthophyll cycle pigments on chlorophyll a fluorescence lifetime distributions and intensity in thylakoids. Biochemistry **37**: 13582-13593.

Golbeck JH, 1992, Structure and function of photosystem I. Annu. Rev. Plant Physiol. Plant Mol. Biol. **43**: 293-324.

Gonzalez EB, Barbato R, Aro EM, 1999, Role of phosphorylation in the repair cycle and oligomeric structure of photosystem II. Planta **208**: 196-204.

Gonzalez-Rodriguez AM, Tausz M, Wonisch A, Jimenez MS, Grill D, Morales D, 2001, The significance of xanthophylls and tocopherols in photo-oxidative stress and photoprotection of three canarian laurel forest tree species on a high radiation day. J. Plant Physiol. **158** : 1547-1554.

Govindjee, 2002, A role for a light-harvesting antennae complex of photosystem II in photoprotection. Plant Cell. **14** : 1663-1668.

Govindjee, Seufferheld MJ, 2002, Non-photochemical quenching of chlorophyll a fluorescence : early history and characterization of two xanthophyll–cycle mutants of *Chlamydomonas reinhardtii* Funct. Plant Biol. **29** : 1141-1155.

Govindjee, Spilotro P, 2002, An *Arabidopsis thaliana* mutant, altered in the r-subunit of ATP synthase has a different pattern of intensity-dependent changes in non-photochemical quenching and kinetics of the P-to-S-fluorescence decay. Funct. Plant Biol. **29**: 425-434.

Grace SC, Logan BA, 2000, Energy dissipation and radical seavenging by the plant phenylpropanoid pathway. Phil. Trans. R. Soc. Lond. B. **355**: 1499-1510.

Grasses T, Grimm B, Koroleva O, 2001, Loss of α-tocopherol in tobacco plants with decreased geranylgeranyl reductase activity does not modify photosynthesis in optimal growth conditions but increases sensitivity to high-light stress. Planta **213**: 620-628.

Grasses T, Pesaresi P, Schiavon F, Varotto C, Salamini F, Jahns P, Leister D, 2002, The role of ΔPH-dependent dissipation of excitation energy in protecting photosystem II against light induced damage in *Arabidopsis thaliana*. Plant Physiol. Biochem. **40**: 41-49.

Gray GR, Savitch LV, Ivanov G, Huner NPA, 1996, Photosystem II excitation pressure and development of resistance to photoinhibition II. Adjustment of photosynthetic capacity in winter wheat and winter rye. Plant Physiol. **110**: 61-71.

Green BR, Durnford DG, 1996, The chlorophyll carotenoid proteins of oxygenic photosynthesis. Annu. Rev. Plant Physiol. Plant Mol. Biol. **47**: 685-714.

Green BR, Salter AH, 1996, Light regulation of nuclear encoded thylakoid proteins: In: B Andersson, AH Salter, J Barber, (eds) Molecular Genetics of Photosynthesis. Oxford University Press, Oxford, pp.75-103.

Greer DH, Berry JA, Bjorkman O, 1986, Photoinhibition of photosynthesis in intact bean leaves : Role of light and temperature and requirement of chloroplast protein synthesis during recovery. Planta **168**: 253-260.

Grotz B, Molnar P, Stransky H, Hager A, 1999, Substrate specificity and functional aspects of violaxanthin de-epoxidase, an enzyme of the xanthophyll cycle. J. Plant Physiol. **154**: 437-446.

Grusak MA, Dellapenna D, 1999, Improving the nutrient composition of plants to enhance human nutrition and health. Annu. Rev. Plant Physiol. Plant Mol. Biol. **50** : 133-161.

Gullner G, Komives T, Rennenberg H, 2001, Enhanced tolerance of transgenic poplar plants overexpresing gama-glutamyl cysteine synthetase towards chloroacetanilide herbicides. J, Exp. Bot. **52** : 971-979.

Hager A, 1969, Lichtbedingte pH- Erniedrigung in einem Chloroplasten-Kompartiment als Ursacheder enzymatischen Violaxanthin–zeaxanthin–Umwandlung; Beziehungen zur Photophos-phorylierung. Planta **89** : 224-243.

Hak R, Lichtenthaler HK, Rinderle U, 1990, Decrease of the fluorescence ratio F 690/F 730 during greening and development of leaves. Rad. Environ. Biophys. **29**: 329-336.

Haldrup A, Jensen PE, Lunde C, Scheller HV, 2001, Balance of power: a view of the mechanism of photosynthetic state transitions. Trends Plant Sci. **6**: 301-305.

Haldrup A, Simson DJ, Scheller HV, 2000, Down – regulation of the PS I–F Sub Unit of Photosystem I in *Arabidopsis thaliana*: The PS I–F subunit is essential for photoautotrophic growth and antennae function. J. Biol. Chem. **275**: 31211-18

Hanba YT, Kogami H, Terashima I, 2002, The effect of growth irradiance on leaf anatomy and photosynthesis in *Acer* species differing in light demand. Plant Cell Environment **25**: 1021-1030.

Hankamer B, Barber J, Boekema EJ, 1997a, Structure and membrane organization of photosystem II in green plants. Annu. Rev. Plant Physiol. Plant Mol. Biol. **48**: 641-671.

Hankamer B, Morris E, Nield J, Carne A, Barber J, 2001, Subunit positioning and transmembrane helix organization in the core dimer of photosystem II. FEBS Lett. **504**: 142-151.

Hankamer B, Morris E, Nield J, Gerle C, Barber J, 2001, Three dimensional structure of photosystem II core dimer of higher plants determined by electron microscopy. J. Struct. Biol. **135**: 262-269.

Hankamer B, Nield J, Zheleva D, Boekema E, Jansson S, Barber J, 1997b, Isolation and biochemical characterisation of monomeric and dimeric photosystem II complexes from spinach and their relevance to the organisation of photosystem II in *vivo*. Eur. J. Biochem. **243**: 422-429.

Hankammer B, Morris EP, Barber J, 1999, Cryoelectron microscopy of photosystem II shows that CP 43 and CP 47 are located on opposite sides of the D1/D2 reaction center proteins. Nature Struct. Biol. **6**: 560-564.

Harrer R, Bassi R, Testi MG, Schafer C, 1998, Nearest neighbour analysis of a PS II complex from *Marchantia polymorpha* (liverwort) which contains reaction center and antennae proteins. Eur. J. Biochem. **255**: 196-205.

Hartel H, Lokstein H, Grimm B, Rank B, 1996, Kinetic studies on the xanthophyll cycle in barley leaves: influence of antenna size and relations to nonphotochemical chlorophyll fluorescence quenching. Plant Physiol. **110** : 471-482.

Haussuhl K, Andersson B, Adamska I, 2001, A chloroplast DegP2 protease performs the primary cleavage of the photodamaged D1 protein in plant photosystem II. EMBO J. (20) **4**: 713-722.

Havaux M, 1998, Carotenoids as membrane stabilizers in chloroplasts. Trends Plant Sci. **3**: 147-151.

Havaux M, Bonfies JP, Lutz C, Niyogi KK, 2000, Photodamage of the photosynthetic apparatus and its dependence on the leaf developmental stage in the npq1 *Arabidopsis* mutant defecient in the xanthophyll cycle enzyme violaxanthein de-epoxidase. Plant Physiol. **124**. 273-284.

Havaux M, Davaud A, 1994, Photoinhibition of photosynthesis in chilled potato leaves is not correlated with a loss of photosystem II activity-preferential inactivation of photosystem I. Photosynth. Res. **40**: 75-92.

Havaux M, Kloppstech K, 2001, The protective functions of carotenoid and flavonoid pigments against excess visible radiation at chilling temperature investigated in *Arabidopsis* npq and tt mutants. Planta **213**: 953-966.

Havaux M, Niyogi KK, 1999, The violaxanthin cycle protects plants from photo-oxidative damage by more than one mehanism. Proc. Natl. Acad. Sci. USA **96**: 8762-8767.

Havaux M, Tardy F, Lemoine Y, 1998, Photosynthetic light harvesting function of carotenoids in higher plant leaves exposed to high light irradiances. Planta **205**: 242-250.

Havir EA, Tausta SL, Peterson RB, 1997, Purification and properties of violaxanthin de-epoxidase from spinach. Plant Sci. **123**: 57-66.

He WZ, Malkin R (1998) Photosystem I and II. In: A S Raghavendra, (ed.) Photosynthesis. Cambridge University Press, U.K., pp. 29-43.

Heber U, Bliguy R, Streb P, Douce R, 1995, Photorespiration is essential for the protection of the photosynthetic apparatus of C3 plants against photoinactivation under sunlight. Bot. Acta **109** : 307-315.

Heber U, Bukhov NG, Shuvalov VA, Kobayashi Y, Langa OL, 2002, Protection of the photosynthetic apparatus against damage by excessive illumination in homoiohydric leaves and poikilohydric mosses and lichens. J. Exp. Bot. **52**: 1999-2006.

Heber U, Walker D, 1992, Concerning a dual function of coupled cyclic electron transport in leaves. Plant Physiol. **100**: 1621-1626.

Henmi T, Yamasaki H, Sakuma S, Tomokawa Y, Tamura N, Shen J-R, Yamamoto Y, 2003, Dynamic interaction between the D1 protein, CP 43 and OEC 33 at the lumenal side of photosystem II in spinach chloroplasts: Evidence from light induced cross–linking of the proteins in the Donor–side photoinhibition. Plant Cell Physiol. **44**: 451-456.

Hieber AD, Bugos RC, Verhoeven AS, Yamamoto HY, 2002, Overexpression of violaxanthin de-epoxidase properties of C terminal deletions on activity and pH-dependent lipid binding. Planta **214**: 476-483.

Hihara Y, Kamei A, Kanehisa M, Kaplan A, Ikeuchi M, 2001, DNA microarray analysis of cyanobacterial gene expression during acclimation to high light. Plant Cell **13**: 793-806.

Hihara Y, Sonoike K, 2001, Regulation, inhibition and protection of photosystem I. In: EM Aro, B Andersson, (eds) Advances in Photosynthesis and Respiration Vol.11. Kluwer Academic Publishers, Dordrecht pp.507-531.

Hihara Y, Sonoike K, Kewchi M, 1998, A novel gene, pmgA, specifically regulates photosystem stoichiometry in the cyanobacterium *Synechocystis* sp. PCC 6803 in response to high light. Plant Physiol. **117**: 1205-1216.

Hillier W, Babcock GT, 2001, Photosynthetic reaction centers. Plant Physiol. **125**: 33-37.

Hirata M, Issh R, Kumara A, Murata Y, 1983, Photoinhibition of photosynthesis in soyabean leaves. II. Leaf orientation – adjusting movement as a possible avoiding mechanism of photoinhibition. Japan J. Crop Science **52**: 319-322.

Hirose T, Sugiura, 1996, Cis-acting elements and trans-acting factors for accurate translation of chloroplast PsbA mRNAs: development of an in vitro translation system from tobacco chloroplasts EMBO J. **15**: 1687-1695.

Hirschberg J, 2001, Carotenoid biosynthesis in flowering plants. Curr. Opin. Plant Biol. **4**: 210-218.

Horton P, Ruban AV, Young JA, 1999, Regulation of the structure and function of the light harvesting complexes of photosystem II by the xanthophyll cycle. In: HA Frank, AJ Young, G Britton, RJ Cogdell, (eds) Advances in Photosynthesis Vol.8. Kluwer Academic Publishers Dordercht, pp. 271-291.

Horton P, Ruban AV, Rees D, Pascal AA, Noctor G, Young AJ, 1991, Control of the light–harvesting function of chloroplast membranes by aggregation of the LHC II chlorophyll – protein complex. FEBS Lett. **292** : 1-4.

Horton P, Ruban AV, Walters RG, 1994, Regulation of light harvesting in green plants. Plant Physiol. **106**: 415-420.

Horton P, Ruban AV, Walters RG, 1996, Regulation of light harvesting in green plants. Annu. Rev. Plant Physiol. Plant Mol. Biol. **47**: 655-684.

Houben E, Gier JW, Van Wijk KJ, 1999, Insertion of leader peptidase into the thylakoid membrane during synthesis in a chloroplast translation system. Plant Cell **11**: 1553-1564.

Howitt CA, Corley JW, Wiskich JT, Vermaas WFJ, 2001, A strain of *Synechocystis* sp.pcc 6803 without photosynthetic oxygen evolution and respiratory oxygen consumption: implications for the study of cyclic photosynthetic clectron transport. Planta **214**: 46-56.

Huber CG, Timperio AM, Zolla L, 2001, Isoforms of photosystem II antennae proteins in different plant species revealed by liquid chromatography – electrospray ionization mass spectroscopy. J. Biol. Chem. **276**(49): 45755-45761.

Huner NPA, Oquist G, Melis A, 2003, Photostasis in plants, green algae and cyanobacteria; the role of light harvesting antennae complexes. In: BR Green, W Parsons, (eds) Advances in Photosynthesis and Respiration. Vol.13. Kluwer Academic Publishers, Dordrecht.

Huner NPA, Oquist G, Sachan F, 1998, Energy balance and acclimation to light and cold. Trends Plant Sci. **3**: 224-230.

Hurry V, Anderson JM, Chow WS, Osmond CB, 1997, Accumulation of zeaxanthin in abscisic acid-deficient mutants of *Arabidopsis* does not affect chlorophyll fluorescence quenching or sensitivity to photoinhibition in vivo. Plant Physiol. **113**: 639-648

Ihalainen JA, Jensen PE, Haldrup A, Van Stokkum IHM, Van Grondelle R, Scheller HV, Dekker JP, 2002, Pigment organization and energy transfer dynamics in isolated photosystem I (PS I) complexes from *Arabidopsis thaliana* depleted of the PSI-G, PSI-K, PSI-L or PSI-N subunit. Biophys. J. **83**: 2190-2201.

Ishikawa Y, Nakatani E, Henmi T, Ferjani A, Harada Y, Tamura N, Yamamoto Y, 1999, Turnover of the aggregates and cross-linked products of the D1 proteins generated by acceptor–side photoinhibition of photolyotene II. Biochim. Biophys. Acta **1413**: 147-158.

Ivanov AG, Sane PV, Zeinalov Y, Malmberg G, Gardestrom P, Huner NPA, Oquist G, 2001, Photosynthetic electron transport adjustments in overwintering scots pine (*Pinus sylvestris* L.). Planta **213**: 575-585.

Ivanov B, Kobayashi Y, Bukhov NG, Heber U, 1998, Photosystem I-dependent cyclic electron flow in intact spinach chloroplasts : Occurrence, dependence on redox conditions and electron acceptors and inhibition by antimycin A. Photosynth. Res. **53**: 61-70.

Ivleva NB, Shestakov SV, Pakrasi H, 2000, The carboxyl terminal extension of the precursor D1 protein of photosystem II is required for optimal photosynthetic performance of the cyanobacterium, *Synechocystis* sp PCC 6803. Plant Physiol. **124**: 1403-1411.

Jackowski G, Jansson S, 1998, Characterization of photosystem II antennae complexes separated by non-denaturing isoelectric fousing. Z. Naturforsch. C. **53**: 841-848.

Jacob B, Heber U, 1994, Photoproduction and detoxification of hydroxyl radicals in chloroplasts and leaves in relation to photoinactivation of photosystems I and II. Plant Cell Physiol. **37** : 629-635.

Jahns P, 1995, The xanthophyll cycle in intermittent light-grown pea plants: Possible functions of chlorophyll a/b-binding proteins. Plant Physiol. **108**: 149-156.

Jahns P, Depka B, Trebst A, 2000, Xanthophyll cycle mutants from *Chlamydomonas reinhardtii* indicate a role for zeaxanthin in the D1 protein turnover. Plant Physiol. Biochem. **38**: 371-376.

Jahns P, Krause GH, 1994, Xanthophyll cycle and energy dependent flourscence quenching in leaves from pea plants grown under intermittent light. Planta **192**: 176-182.

Jansson S, 1994, The light harvesting chlorophyll a/b binding proteins. Biochim. Biophys. Acta **1184**: 1-19.

Jansson S, 1999, A guide to the Lhc genes and their relatives in *Arabidopsis*. Trends Plant Sci, **4**: 236-240.

Jansson S, Pichersky E, Bassi R, Green BR, Ikeneti M, Molis A, Simpson DJ, Spangfort M, Stachelis LA, Thornber JP, 1992, A nomenclature for the genes encoding the chlorophyll a/b binding proteins of higher plants. Plant Mol. Biol. Rep. **10**: 242-253.

Jarvis PG, James BG, Landsberg JJ, 1976, Coniferous forests. In: JL Monteith, (ed) Vegetation and Atmosphere. Academic Press, London pp.171-204.

Jarvis PG, Stanford AP, 1986, Temperate forests. In: NR Baker, SP Long, (eds) Photosynthesis in Contrasting Environments. Elsevier, Oxford, pp.200-228

Jegerschold C, Virgin I, Styring S, 1990, Light dependent degradation of the D1 protein in photosystem II is acclerated after inhibition of the water – splitting reaction. Biochemistry **29**: 6179-6186.

Jensen PK, Rosgaard L, Knoetzel J, Scheller HV, 2002, Photosystem I activity is increased in the absence of the PS I-G subunit. J. Biol. Chem. **277**: 2798-2803.

Jeong WJ, Park Y, Suh K, Raven JA, Yoo OJ, Liu JR, 2002, A large population of small chloroplasts in tobacco leaf cells allows more effective chloroplast movement than a few enlarged chloroplasts. Plant Physiol. **129**: 112-121.

Jin E, Yokthangwattana K, Polle JEW, Melis A, 2003, Role of the reversible xanthophyll cycle in the photosystem II damage and repair cycle in *Dunaliella salina*. Plant Physiol. **132**: 352-364.

Joet T, Cournae L, Peltier G, Havaux M, 2002, Cyclic electron flow around photosystem I in C3 plants. In vivo, control by the redox state of chloroplasts and involvement of the NADH–dehydrogenase complex. Plant Physiol. **128**: 760-769.

Joet T, Cournae L, Horvath EM, Medgyesy P, Peltier G, 2001, Increased sensitivity of photosynthesis to antimycin A induced by inactivation of the chloroplast ndhB gene. Evidence for a participation of the NADH–dehydrogenase complex to cyclic electron flow around photosystem I. Plant Physiol. **125**: 1919-1929.

Jordan P, Fromme P, Witt HT, Klukas O, Saenger W, Krauss N, 2001, Three dimensional structure of cyanobacterial photosystem I at 2.5 Å resolution. Nature. **411**: 909-917.

Jurik TW, 1986, Temporal and spetial patterns of specific leaf weight in successional northem hard wood tree species. Am. J. Bot. **73**: 1083-1092.

Kagawa T, Wada M, 2002, Blue light – induced chloroplast relocation. Plant Cell Physiol. **43**: 367-371.

Kamiya N, Shen JR, 2003, Crystal structure of oxygen evolving photosystem II from *Thermosynechococcus vulcanus* at 3.7- Å resolution. Proc. Natl. Acad. Sci. USA **100**: 98-103.

Kaneko T, Tabata S, 2001, Functional genomics in *Synechocystis* sp. PCC 6803: resources for comprehensive studies of gene function and regulation. In: EM Aro, B Andersson, (eds) Advances in Photosynthesis Vol.11 Kluwer Academic Publishers Dordrecht pp.557-561.

Kao W, Forseth IN, 1992, Diurnal leaf movement, chlorophyll fluorescence and carbon assimilation in soyabean grown under different nitrogen and water availabilities. Plant Cell Environ. **15**: 703-710.

Kao WY, Comstock JP, Ehleringer JR, 1994, Variation in leaf movements among common bean cultivars. Crop Sci. **34**: 1273-1278.

Kao WY, Tsai TT, 1998, Tropic leaf movements, photosynthetic gas exchange, leaf ^{13}C and chlorophyll a fluorescence of three soyabean species in response to water availability. Plant Cell Environ. **21**: 1055-1062.

Karpinski S, Escobar C, Karpinska B, Creissen G, Mullineaux, 1997, Photosynthetic electron transport regulates the expression of cytosolic ascorbate peroxidase genes in *Arabidopsis* during excess light stress. Plant Cell **9**: 627-640.

Karpinski S, Reynolds H, Karpinska B, Wingsle G, Creissen G, Mullineaux P, 1999, Systemic signaling and acclimation in response to excess excitation energy in *Arabidopsis*. Science **284**: 654-657.

Karpinski S, Wingsle G, Karpinska B, Hallgren JE, 2001, Redox sensing of photo-oxidative stress and acclimatory mechanisms in plants. In: EM Aro, B Andersson, (eds) Advances in Photosynthesis and Respiration. Vol.11 Kluwer Academic Publishers, Dordrecht pp.469-486.

Kasahara M, Kagawa T, Oikawa K, Suetsugu N, Miyao M, Wada M, 2002, Chloroplast avoidance movement reduces photodamage in plants. Nature **420** : 829-832.

Kato M, Hikosaka K, Hirotsu N, Makino A, Hirose T, 2003, The excess light energy that is neither utilized in photosynthesis nor dissipated by photoprotective mechanisms determines the rate of photoinactivation in photosystem II. Plant Cell Physiol. **44**: 316-325.

Kato MC, Mikosaka K, Hirose T, 2002, Photoinactivation and recovery of photosystem II in *Chenopodium album* leaves grown at different levels of irradiance and nitrogen availability. Funct. Plant Biol. **29**: 787-795.

Ke B, 2000a, Role of carotenoids in photosynthesis. In: Govindjee, (ed.) Advances in Photosynthesis, Vol.10 Kluwer Academic Publishers, Dordrecht. pp.229-250.

Ke B, 2000b, The stable primary electron acceptor Q_A and the secondary electron acceptor Q_B. In: Goindjee, (ed.) Advances in Photosynthesis, Vol.10 Kluwer Academic Publishers, Dordrecht pp.289-304.

Keegstra K, Cline K, 1999, Protein import and routing systems of chloroplasts. Plant Cell **11**: 557-570.

Kettunen R, Tyystjarvi E, Aro EM, 1996, Degradation pattern of photosystem II reaction center protein D1 in intact leaves: The major photoinhibition – induced cleavage site on D1 polypeptide is located aminoterminally of the DE loop. Plant Physiol. **111**: 1183-1190.

Khorobrykh SA, and Ivanav BN, 2002, Oxygen reduction in plastoquinone pool of isolated pea thylakoids. Photosynth. Res. **71**: 209-219.

Kim J, Mayfield SP, 1997, Protein disulphide isomerase as a regulator of chloroplast translational activation. Science **278**: 1954-1957.

Kim JH, Melis A, 1992, Mechanism of chloroplast acclimation to subtropical and adverse irradiance. In: N.Murata, (ed.) Research in Photosynthesis Vol. 14, Kluwer Academic Publishers, Netherlands, pp.317-324.

Kimura M, Yamamoto Y, Seki M, Sakurai T, Sato M, Abe T, Yoshida S, Manabe K, Shimozaki K, Matsui M, 2003, Identification of *Arabidopsis* genes regulated by high light–stress using cDNA microarray. Photochem. Photobiol. **77**: 226.

Kitao M, Lei LT, Koike T, Tobita H, Maruyama, 2000, Susceptibility to photoinhibition of three deciduous broad leaf tree species with different successional traits raised under various light regimes. Plant Cell Environ. **23**: 81-89.

Knoetzel J, Svendsen I, Simpson DJ, 1992, Identification of the Photosystem I antennae polypeptides in barley. Isolation of three pigment binding antennae complexes. Eur. J. Biochem. **206**: 209-215.

Kogata N, Nishio K, Hirohashi T, Kikuchi S, Nakai M, 1999, Involvement of a chloroplast homologue of the signal recognition particle receptor protein, FtsY, in protein targeting to thylakoids. FEBS Lett **329**: 329-333.

Kok B, 1956, On the inhibition of photosynthesis by intense light. Biochim. Biophys. Acta **21**: 234-244

Koller D, 1990, Light driven leaf movements. Plant Cell Environ. **13**: 615-632.

Koller D, 2000, Plants in search of sunlight. Adv. Bot. Res. **33**: 35-131

Koller D, Ritter S, 1994, Phototropic responses of the pulvinules and associated laminar reorientation in the trifoliate leaf of bean *Phaseolus vulgaris* (Fabaceae). J. Plant Physiol. **143**: 52-63.

Konlougliotis D, Innes JB, Brudvig GW, 1994, Location of chlorophyll z in photosystem II. Biochemistry **33**: 11814-11822.

Kozaki A, Takeba G, 1996, Photorespiration protects C3 photosynthesis from photo-oxidation Nature. **384**: 557-560.

Krause GH, 1988, Photoinhibition of photosynthesis. An evaluation of damaging and protecting mechanisms Physiol. Plant **74**: 566-574.

Krause GH, 1994, Photoinhibition induced by low temperatures. In: NR Baker, JR Bowyer, eds, Photoinhibition of Photosynthesis. From Molecular Mechanisms to Field. BIOS Scientific Publishers, Oxford, pp.331-348.

Krause GH, Brintias J-M, Vernotte C., 1983, Characterisation of chlorophyll fluorescence quenching in chloroplasts by fluorescence spectroscopy at 77K I. Δ pH – dependent quenching. Biochim. Biophys. Acta **723**: 169-175.

Krause GH, Wiese E, 1991, Chlorophyll fluorescence and photosynthesis: the basics. Annu. Rev. Plant Physiol. Plant Mol. Biol. **42**: 313-349.

Krauss N, Sehabert W-D, Klukas O, Fromme P, Witt HT, Senger W, 1996, Photosystem I at 4 Å resolution represents the first structural model of a joint photosynthetic reaction center and core antennae system. Nature Struct. Biol. **3**: 965-973.

Krieger A, Moya I, Weis E, 1992, Energy – dependent quenching of chlorophyll a fluorescence: effect of pH on stationary fluorescence and picosecond relaxation kinetics in thylakoid membranes and photosystem II preparations. Biochim. Biophys. Acta **1102**: 167-176.

Krol M, Spangfort MD, Huner NPA, Oquist G, Gustaffson P, Jansson S, 1995, Chlorophyll a/b – binding proteins, pigment conversions and early light – induced proteins in a chlorphyll b-less barley mutant. Plant Physiol. **70**: 1242-1248.

Kuhlbrandt W, 2001, Chlorophylls galore. Nature. **411**: 896-899.

Kuhlbrandt W, Wang DN, Fujiyoshi Y, 1994, Atomic model of plant light – harvesting complex Nature **367**: 614-621.

Kulheim C, Ågren J, Jansson S, 2002, Rapid regulation of light harvesting and plant fitness in the field. Science **297**: 91-93.

Kursar TA, Coley PD, 1999, Contrasting modes of light acclimation in two species of rainforest understory. Oecologia **121**: 489-498.

Kuwabara T, Hasegawa M, Kawano M, Takaichi S, 1999, Characterisation of violaxanthin de-epoxidase purified in the presence of Tween 20 : Effects of dithiothreitol and pepstatin A. Plant Cell Physiol. **40**: 1119-1126.

Lagoutte B, Hanley J, Bottin H, 2001, Multiple functions for the C terminus of the PsaD subunit in the cyanobacterial Photosystem I Complex. Plant Physiol. **126**: 307-316.

Lancaster CRD, Bibikova MV, Sabatino P, Oesterhelt D, Michel H, 2000, Structural basis of the drastically increased initial electron transfer rate in the reaction center from a *Rhodopseudomonas viridis* mutant described at 2.00 Å resolution. J. Biol. Chem. **275**: 39364-39368.

Lawson T, Oxborough K, Morison JIL, Baker NR, 2002, Responses of photosynthetic electron transport in stomatal guard cells and mesophyll cells in intact leaves to light, CO_2 and humidity. Plant Physiol. **128**: 52-62.

Lazar D, 1999, Chlorophyll a fluorescence induction. Biochim. Biophys. Acta **1412** : 1-28.

Lee HY, Chow WS, Hong YN, 1999, Photoinactivation of photosystem II in leaves of *Capsicum annuum*. Physiol. Plant **105**: 377-384.

Lee HY, Hong YN, Chow WS, 2001, Photoinactivation of photosystem II complexes and photoprotection by non-functional neighbours in *Capsicum annuum* L. Leaves Planta **212**: 332-342.

Lee HY, Hong YN, Chow WS, 2002, Putative effects of pH in intra-chloroplast compartments on photoprotection of functional photosystem II complexes by photoinactivated neighbours and on recovery from photoinactivation in *Capsicum annuum* leaves. Funct. Plant Biol. **29**: 607-619.

Leong TY, Anderson JM, 1984, Adaptation of the thylakoid membrane of pea chloroplasts to light intensities II. Regulation of electron transport capacities, electron transport carriers, coupling factor (CF1) activity and rates of photosynthesis. Photosynth. Res. **5**: 117-128.

Levy I, Gnatt E, 1988, Light acclimation in *Porphyridium purpurium* (Rhodophyta) growth, photosynthesis and phycobilisomes. J. Phycol. **24**: 488-495.

Li XP, Bjorkman O, Shih C, Grossman AR, Rosenquist M, Jansson S, Niyogi K, 2000, A pigment–binding protein essential for regulation of photosynthetic light harvesting. Nature **403**: 391-395.

Li XP, Miiller-Moule P, Gilmore AM, Niyogi KK, 2002, PsbS – dependent enhancement of feedback de-exicitation protects photosystem II from photoinhibition. Proc. Natl. Acad. Sci. USA **99**: 15222-15227.

Lichtenthaler HK, Roman H, Rinderle U, 1990, The chlorophyll fluorescence ratio F 690/F 730 in leaves of different chlorophyll content. Photosynth. Res. **25**: 295-298.

Lindahl M, Spetea C, Hundal T, Oppenheim A, Adam Z, Andersson B, 2000, The thylakoid FtsH protease plays a role in the light induced turnover of the photosystem II D1 protein. Plant Cell **12**: 419-431.

Lindahl M, Tabak S, Eseke L, Pichersky E, Andersson B, Adam Z, 1996, Identification, characterization and molecular cloning of a homologue of the bacterial FtsH protease in chloroplasts of higher plants. J. Biol. Chem. **271**: 29329-29334.

Lindahl M, Yang D-H, Andersson B, 1995, Regulatory proteolysis of the major light–harvesting chlorophyll a/b protein of photosystem II by a light–induced membrane–associated enzymic system. Eur. J. Biochem. **231**: 503-509.

Logan A, Demmig-Adams B, Adams III WW, Grace SC, 1998, Antioxidants and xanthophyll cycle dependent energy dissipation in *Cucurbita pepo* L. and *Vinca major* L. acclimated to four growth PPFDs in the field. J. Exp. Bot. **49**: 1869-1879.

Logan BA, Grace SC, Adams III WW, 1998, Seasonal differences in xanthophyll cycle characteristics and antioxidants in *Mahonia repens* growing in different light environments. Oecologia **116**: 9-17.

Lokstein H, Hartel H, Hoffmann P, Woitke P, Renger G, 1994, The role of light-harvesting complex II in excess excitation energy dissipation: An in vivo fluorescence study on the origin of high-energy quenching, J. Photochem. Pholophiol. **26**: 175-184.

Lokstein M, Tian L, Polle JEW, Della Penna D, 2002, Xanthophyll biosynthetic mutants of *Arabidopsis thaliana*: Altered nonphotochemical quenching of chlorophyll fluoresence is due to changes in photosystem II antennae size and stability. Biochim. Biophys. Acta **1553**: 309-319.

Long SP, Humphries S, Falkowski PG, 1994, Photoinhibition of photosynthesis in nature. Annu. Rev. Plant Physiol. Plant Mol. Biol. **45**: 633-662.

Maenpaa P, Andersson B, Sundby C, 1987, Difference in sensitivity to photoinhibition between photosystem II in the appressed and non-appressed thylakoid regions. FEBS Lett. **215**: 31-36.

Marin E, Nussaume L, Quesada A, Gonneau M, Sotta B, Huguency P, Frey A, Marion–Poll A, 1996, Molecular identification of zeaxanthin epoxidase of *Nicotiana plumbaginifolia*, a gene involved in abscisic acid biosynthesis and corresponding to the ABA locus of *Arabidopsis thaliana* . EMBO J. **15**: 2331-2342.

Marshall HL, Geides RJ, Flynn K, 2000, A mechanistic model of photoinhibition. New Phytol. **145**: 347-359.

Masuda T, Polle JEW, Melis A, 2002, Biosynthesis and distribution of chlorophyll among the photosystems during recovery of the green algae *Dunaliella salina* from irradiance stress. Plant Physiol. **128**: 603-614.

Maxwell K, Johnson GN, 2000, Chlorophyll fluorescence. A practical guide. J. Exp. Bot. **51**: 659-668.

Maxwell K, Marrison JL, Leech RM, Griffiths H, Horton P, 1999, Chloroplast acclimation in leaves of *Guzmania monostachia* in response to high light. Plant Physiol. **121**: 89-95.

Mehler AH, 1951, Studies on the reactions of illuminated chloroplasts I: Mechanism of the reduction of oxygen and other Hill reagents. Arch. Biochem. Biophys. **33**: 65-77.

Meir P, Kruijt B, Broadmeadow M, Barbosa E, Kull O, Carswell F, Nobre A, Jarvis PG, 2002, Acclimation of photosynthetic capacity to irradiance in tree canopies in relation to leaf nitrogen concentration and leaf mass per unit area. Plant Cell Environ. **25**: 343-357.

Melis A,1991, Dynamics of photosynthetic membrane composition and function. Biochim.Biophys. Acta **1058**: 87-106.

Melis A, 1996, Excitation energy transfer, functional and dynamic aspects of LHC (cab) proteins. In: DR Ort, CF Yocum, (eds) Advances in Photosynthesis Vol.4 Kluwer Academic Publishers Netherlands. pp.523-538.

Melis A, 1999, Photosystem II damage and repair cycle in chloroplasts: what modulates the rate of photodamage in vivo. Trends Plant Sci. **4**: 130-135.

Melis A, Manodori A, Glick RE, Ghirardi ML, McCauley SW, Neale PJ, 1985, The mechanism of photosynthetic membrane adaption to environmental stress conditions: A Hypothesis on the role of electron transport and of ATP/NADPH pool in the regulation of thylakoid membrane organization and function. Physiol. Veg. **23**: 757-765.

Meyer AJ, Fricker MD, 2002, Control of demand – driven biosynthesis of glutathione in green *Arabidopsis* suspension culture cells. Plant Physiol. **130**: 1927-1937.

Meyer WS, Walker S, 1981, Leaflet orientation in water-stressed soyabeans. Agron. J. **73**: 1071-1074.

Michelet B, Boutry M, 1995, The plasma membrane H^+-ATPase A highly regulated enzyme with multiple physiological functions. Plant Physiol. **108**: 1-6.

Miyake C, Okamura M, 2003, Cyclic electron flow within PS II protects from its photoinhibition in thylakoid membranes from spinach chloroplasts. Plant Cell Physiol. **44**: 457-462.

Miyake E, Yokota A, 2000, Determination of the rate of photoreduction of O_2 in the water-water cycle in watermelon leaves and enhancement of the rate by limitation of photosynthesis. Plant Cell Physiol. **41** (3): 335-343.

Mo F, 1995, X-ray crystallographic studies. In: Britton S, Liaaen– Jenson, H Pfander, (eds) Carotenoids Vol.I B Spectroscopy. Birkhauser, Basel. pp.321-342.

Morais F, Barber J, Nixon PJ, 1999, The chloroplast – encoded subunit of cytochrome b559 is required for assembly of photosystem II complex in both the light and dark in *Chlamydomonas reinhardii*. J. Biol. Chem. **273**: 29315-29320.

Morishige DT, Dreyfuss BW, 1998, Light harvesting complexes of higher plants. In: AS Raghavendra, (ed.) Photosynthesis, Cambridge Univ Press. pp.18-27.

Muller P, Li XP, Niyogi KK, 2001, Non-photochemical quenching. A. response to excess light energy. Plant Physiol. **125**: 1358-1366.

Mullineaux P, Ball L, Escobar C, Karpinska B, Creissen G, Karpinski S, 2000, Are diverse signaling pathways integrated in the regulation of excess exitation energy? Phil. Trans. R. Soc. Lon. B **355**: 1531-1540.

Mullineaux P, Karpinski S, 2002, Signal transduction in response to excess light: Getting out of the chloroplast. Curr. Opin. Plant Biol. **5**: 43-48.

Mulo P, Laakso S, Maenpaa P, Aro EM, 1998, Stepwise photoinhibition of photosystem II. Plant Physiol. **117**: 483-490.

Munekage Y, Takeda S, Endo T, Jahns P, Hashimoto T, Shikanai T, (2001) Cytochrome b_6f mutation specifically affects thermal dissipation of absorbed light energy in *Arabidopsis*. Plant J. **28**: 351-359.

Murakami A, Fujita Y, 1991, Regulation of photosystem stoichiometry in the photosynthetic system of the cyanophyte *Synechocystis* PCC 6714 in response to light – intensity. Plant Cell Physiol. **32**: 223-230.

Murchie E, Horton P, 1998, Contrasting patterns of photosynthetic acclimation to the light environment are dependent on the differential expression of the responses to altered irradiance and spectral quality. Plant Cell Environ. **21**: 139-148

Murchie EH, Chen YZ, Hubbart S, Deng S, Horton P, 1999, Interactions between senescence and leaf orientation determine in situ patterns of photosynthesis and photoinhibition in field grown rice. Plant Physiol. **119**: 553-563.

Murchie EH, Hubbart S, Chen Y, Peng S, Horton P, 2002, Acclimation of rice photosynthesis to irradiance under field conditions. Plant Physiol. **130**: 1999-2010.

Murgia I, Briat JF, Tarantino D, and Soave C, 2001, Plant ferritin accumulates in response to photoinhibition but its ectopic over expression does not protect against photoinhibition. Plant Physiol. Biochem. **39**: 797-805.

Nanba O, Satoh K, 1987, Isolation of photosystem II Haction Center consisting of D1 and D2 polypeptides and cytochrome b 559. Proc. Natl. Acad. Sci. USA **84**: 109-112.

Naver H, Haldrup A, Scheller HV, 1999, Cosuppression of photosystem I subunit PS I–H in *Arabidopsis thaliana*. Efficient electron transfer and stability of photosystem I is dependent upon PS I–H subunit. J. Biol. Chem. **274**: 10784-10789.

Nedbal L, Brezina V, 2002, Complex metabolic oscillations in plants forced by harmonic irradiance. Biophys. J. **83**: 2180-2189.

Neubauer C, 1993,Multiple effects of dithiothreitol on nonphotochemical fluorescence quenching in intact chloroplasts. Plant Physiol.**103**: 575-583.

Nield J, Funk C, Barber J, 2000, Super molecular structure of photosystem II and location of the PsbS protein. Phil. Trans. R. Soc. Lond. B. **355**: 1337-1344.

Nield J, Kruse O, Ruprecht J, Fonseca PD, Biiehel C, Barber J, 2000, Three dimensional structure of *Chlamydomonoas reinhardtii* and *Synechococcus elongatus* photosystem II complexes allows for comparison of their oxygen evolving complex organization. J. Biol. Chem. **275**: 27940-27946.

Nilson R, Brunner J, Hoffman NE, Van Wijk KJ, 1999, Interaction of ribosome nascent chain complexes of the chloroplast – encoded D1 thylakoid membrane protein with CPSRP 54. EMBO J **18**: 733-742.

Nishikami M, Yagi K, 1996, Biochemistry and molecular biology of ascorbic acid biosynthesis. In: J Harris, (ed.) Subcellular Biochemistry of Ascorbic acid: Biochemistry and Biomedial Cell Biology. Plenum Press, New York, pp.17-39.

Nixon PJ, 2000, Chlororespiration. Phil. Trans. R. Soc. Lond. B. **355**: 1541-1547.

Nixon PJ, Mullineaux CW, 2001, Regulation of photosynthetic electron transport. In: EM Aro, B Andersson, (eds) Advances in Photosynthesis and Respiration. Vol.11. Kluwer Academic Publishers, Dordrecht, Netherlands, pp.533-555.

Niyogi KK, Grossman AR, and Bjorkman O, 1998, *Arabidopsis* mutants define a central role for the xanthophyll cycle in the regulation of photosynthetic energy conversion. Plant Cell, **10**: 1121-1134.

Niyogi KK, 1999, Photoprotection revisited: genetic and moleculer approaches. Ann. Rev. Plant. Physiol. Plant Mol. Biol. **50**: 333-358.

Niyogi KK, 2000, Safety values for photosynthesis. Curr. Opin. Plant Biol. **3**: 455-460.

Niyogi KK, Bjorkman O, Grassman AR, 1997, The roles of specific xanthophylls in photoprotection. Proc. Natl. Acad. Sci. USA **94**: 14162-14167.

Niyogi KK, Grossman AR, Bjorkman O, 1998, *Arabidopsis* mutants define a central role for the xanthophyll cycle in the regulation of photosynthetic energy conversion. Plant Cell **10**: 1121-1134.

Niyogi KK, Shih C, Chow WS, Pogson BJ, Della Penna D, Bjorkman O, 2001, Photoprotection in a zeaxanthin and lutein deficient double mutant of *Arabidopsis*. Photosynth. Res. **67**: 139-145.

Noctor G, Foyer CH, 1998, Ascorbate and Glutathione. Keeping active oxygen under control. Ann. Rev. Plant Physiol. Plant Mol. Biol. **49**: 249-279.

Noctor G, Horton P, 1990, Uncoupler titration of energy – dependant chlorophyll fluorescence and photosystem II photochemical yield in intact pea chloroplasts. Biochim. Biophys. Acta **1016**: 228-234.

Obokata J, Mikami K, Hayashida N, Nakamura M, Sugiura M, 1993, Molecular heterogeneity of photosystem I. PsaD, PsaE, PsaF, PsaH and PsaL are all present in isoforms in *Nicotiana* spp. Plant Physiol. **102**: 1259-1267.

Ogren E, 1994, The significance of photoinhibition for photosynthetic productivity. In: NR Baker, JR Bowyer, (eds) Photoinhibition of Photosynthesis. Bios Scientific Publishers, Oxford, pp.432-446.

Oguchi R, Hikosaka K, Hirose T, 2003, Does the photosynthetic light – acclimation need change in leaf anatomy? Plant Cell Environ. **26**: 505-512.

Ohad I, Kyle DJ, Arntzen CJ, 1984, Membrane protein damage and repair removal and replacement of inactivated 32.kDa polypeptide in chloroplast membranes. J. Cell Biol. **99**: 481-485.

Ohad I, Sonoike K, Anderson B, 2000, Photoinactivation of the two photosystems in oxygenic photosynthesis: mechanisms and regulations. In: M Yunus, V Pathre, P Mohanty, (eds) Probing Photosynthesis. Taylor and Francis, London, pp.293-309.

Ohnishi N, Takahaski Y, 2001, PsbT Polypeptide is required for efficient repair of photodamaged photosystem II reaction center. J. Biol. Chem. **276**(36): 33798-33804.

Oosterhuis DM, Walker S, Eastham J, 1985, Soyabean (*Glycine max*) leaflet movements as an indicator of crop water stress. Crop Sci. **25**: 1101-1106.

Oquist G, Chow WS, Anderson JM, 1992, Photoinactivation of photosynthesis represents a mechanism for the long term regulation of photosystem II. Planta **186**: 450-460.

Ort DR, 2001, When there is too much light. Plant Physiol. **125**: 29-32.

Ort DR, Baker NR, 2002, A photoprotective role for O_2 as an alternative electron sink in photosynthesis. Curr. Opin. Plant Biol. **5**: 193-198.

Ortega JM, Roncel M, Losada M, 1999, Light-induced degradation of cytochrome b559 during photoinhibition of the photosystem II reaction center. FEBS Lett. **458**: 87-92.

Osmond B, Badgers M, Maxwell K, Bjorkman O, Leegood R, 1997, Too many photons. Photorespiration, photoinhibition and photo-oxidation. Trends Plant Sci. **2**: 119-121.

Osmond CB, 1994, What is photoinhibition: some insights from comparisons of shade and sun plants. In: NR Baker, JR Bowyer eds, Photoinhibition of Photosynthesis. Bios Scientific Publishers, Oxford pp.1-24.

Osmond CB and Grace SC, 1995, Perspectives on photoinhibition and photorespiration in the field: Quintessential inefficiencies of the light and dark reactions of photosynthesis? J. Exp. Bot. **48**: 1351-1362.

Oswald O, Martin T, Dominy PJ, Graham IA, 2001, Plastid redox state and sugars: interactive regulators of nuclear-encoded photosynthetic gene expression. Proc. Natl. Acad. Sci. USA **98**: 2047-2052.

Ott T, Clarke J, Birks K, Johnson G, 1999, Regulation of the photosynthetic electron transport chain. Planta **209**: 250-258.

Owens TG, 1994, Excitation energy transfer between chlorophylls and carotenoids. Proposed molecular mechanism for non photochemical quenching. In: NR Baker and JR Bowyer, (eds) Photoinhibition of Photosynthesis — From Molecular Mechanisms to the Field. Bios Scientific Publishers, Oxford, pp.95-109.

Owens TG, 1996, Processing of excitation energy by antennae pigments. In: NR Baker, (ed) Photosynthesis and the Environment, Kluwer Academic Publishers, The Netherlands. pp.1-23.

Owens TG, Shreve AP, Albreht AC, 1992, Dynamics and mechanism of singlet energy transfer between carotenoids and chlorophylls. Light harvesting and non photochemical flourscence quenching. In: N Murata, (ed) Research in Photosynthesis Vol.I: Kluwer Academic Publishers, Dordrecht pp.179-186.

Palatnik J, Carrillo N, Valle EM, 1999, The role of photosynthetic electron transport in the oxidative degradation of chloroplastic glutamine synthetase. Plant Physiol. **121**: 471-478.

Palmer JM, Warpeha KMF, Briggs WR, 1996, Evidence that zeaxanthin is not the photoreceptor for phototropism in maize coleoptiles. Plant Physiol. **110**: 1323-1328.

Park Yll, Chow WS, Anderson JM, 1997, Antennae size dependences of photoinactivation of photosystem II in light-acclimated pea leaves. Plant Physiol. **115**: 151-157.

Paulsen H, 1995, Chlorophyll a/b – binding proteins. Photochem. Photobiol. **62**: 367-382.

Peltier G, Cournac L, 2002, Chlororespiration. Annu. Rev. Plant. Biol. **53**: 523-550.

Peltier JB, Emanuelsson O, Kalume DE, Ytterberg J, Friso G, Rudella A, Liberles DA, Soderberg L, Roepstorff P, Von Heijne G, Uan Wijk KJ, 2002, Central functions of the lumenal and peipheral thylakoid proteome of *Arabidopsis* determined by experimentation and genome-wide prediction. Plant Cell **14**: 211-236.

Pesaresi P, Varotto C, Richly E, Kurth J, Salamini F, Leister D, 2001, Functional genomics of *Arabidopsis* photosynthesis Plant Physiol. Biochem. **39**: 285-294.

Peterson RB, Havir EA, 2001, Photosynthetic properties of an *Arabidopsis thaliana* mutant possessing a defective PsbS gene. Planta. **214**: 142-152.

Pfannschmidt T, Allen JF, Oelmuller R, 2001a, Principles of redox control in photosynthesis gene expression. Physiol. Plant **112**: 1-9.

Pfannschmidt T, Schutze K, Brost M, Oelmueller R, 2001b, A novel mechanism of nuclear photosynthesis gene regulation by redox signals from the chloroplast during photosystem stoichiometry adjustment. J. Biol. Chem. **276**: 36125-36130.

Pfanschimdt T, Nilssion A, Allen JF, 1999, Photosynthetic Control of chloroplast gene expression. Nature **397**: 625-628.

Pfundel EE, Bilger W, 1994, Regulation and Possible function of the violaxanthin cycle. Photosynth. Res. **42**: 89-109.

Phillip Y, Young AJ, 1995, Occurrence of the carotenoid lactucaxanthin in higher plant LHC II. Photosynth. Res. **43**: 273-282.

Pichersky E, Jansson S, 1996, The light harvesting chlorophyll a/b binding polypeptides and their genes in angiosperm and gymnosperm species. In: DR Ort and CF Yocum (eds) Advances in Photosynthesis, Vol.4. Kluwer Academic Publishers, Netherlands, pp.507-521.

Pineau B, Gerard-Hirne C, Selve C, 2001, Carotenoid binding to photosystem I and II of *Chlamydomonas reinhandtii* cells grown under weak light or exposed to intense light. Plant Physiol. Biochem. **39**: 73-85.

Pogson BJ, Niyogi KK, Bjorkman O, Dellapenna D, 1998, Altered xanthophyll compositions adversely affect chlorophyll accumulation and non-photochemical quenching in *Arabidopsis* mutants. Proc. Natl. Acad. Sci. USA. **95**: 13324-13329.

Pogson H, Mcdonald KA, Truong M, Britton G, Dellapenna D, 1996, *Arabidopsis* carotenoid mutants demonstrate that lutein is not essential for photosynthesis in higher plants. Plant Cell **8**: 1627-1639.

Polle A, 1996, Mehler reaction : Friend or foe in photosynthesis? Bot. Acta **109**: 84-89.

Polle A, 1997, Defense against photo-oxidative damage in plants. In: JG Scandalios, (ed) Oxidative stress and Molecular biology of antioxidant defenses. Cold Spring Harbor Laboratory Press. Cold Spring Harbor, New York, pp.623-666.

Polle A, 2001, Dissecting the superoxide dismutase – ascorbate – glutathione – pathway in chloroplasts by metabolic modelling. Computer simulations as a step towards flux analysis. Plant Physiol. **126**: 445-462.

Powles SB, 1984, Photoinhibition of photosynthesis induced by visible light. Annu. Rev. Plant. Physiol. **35**: 15-44.

Prasad JSR, Das VSR, 1984, *Cleome pilosa* Benth: A C_3 plant with high photosynthetic efficiency and solar tracking ability. Curr. Sci. **53**: 1100-1101.

Prasil O, Adir N, Ohad I, 1992, Dynamics of photosystem II : Mechanisms of photoinhibition and recovery processes. In: J Barber, (ed) Topics in photosynthesis, Structure, Function and Molecular Biology. Elsevier Science Publishers, Amsterdam Vol.**11**, pp.295-348.

Psylinakis E, Fritzsch G, Ghanotakis DF, 2002, Isolation and crystallization of CP47, a photosystem II chlorophyll binding protein. Degradation of CP47 upon dissociation from the core complex. Photosynth. Res. **72**: 211-216.

Quigg A, Beardall J, Wydrzynski T, 2003, Photoacclimation involves modulation of the photosynthetic oxygen–evolving reactions in *Dunaliella tertiolecta* and *Phaeodactylum tricornutum*. Funct. Plant Biol. **30**: 301-308.

Raines CA, Lloyd JC, 1996, Molecular biological approaches to environmental effects on photosynthesis. In: NR Baker, (ed) Photosynthesis and the Environment. Kluwer Academic Publishers, Netherlands. pp.305-319.

Rajendrudu G, Das VSR, 1981, Solar tracking and light interception by leaves of some dicot species. Curr. Sci. **50**: 618-620.

Reddy VC, Raghavendra AS, Das VSR, 1983, Photosynthetic units and carbon assimilation in leaves of grain sorghum under different light intensities. Plant Cell Physiol. **24**: 1395-1400.

Rees D, Noctor GD, Horton P, 1990, The effect of high-energy state excitation quenching on maximum and dark level fluorscence yield. Photosynth. Res. **31**: 199-211.

Rhee K, 2001, Photosystem II the solid structural era. Annu. Rev. Biophys. Biomol. Struct. **30**: 307-328.

Rhee KH, Morris EP, Barber J, Kuhlbrandt W, 1998, Three dimensional structure of the plant photosystem II reaction center at 8Å resolution. Nature **396**: 283-286.

Richter M, Goss R, Wagner M, Holzwarth AR, 1999, Characterisation of the fast and slow reversible components of nonphotochemical quenching in isolated pea thylakoids by picosecond time-resolved chlorophyll fluorescence analysis. Biochemistry **38**: 12718-12726.

Rissler HM, Pogson BJ, 2001, Antisense inhibition of the beta-carotene hydroxylase enzyme in *Arabidopsis* and the implications for carotenoid accumulation, photoprotection and antennae assembly. Photosynth. Res. **67**: 127-137.

Robinson C, Hynds PJ, Robindon D, Mant A, 1998, Multiple pathways for the targeting of thylakoid proteins in chloroplasts. Plant Mol Biol **38**: 209-221.

Robinson C, Mant A, 1997, Targeting of proteins into and across the thylakoid membrane. Trends Plant Sci. **2**: 431-437.

Rochaix JD, 2001a, Assembly function and dynamics of the photosynthetic machinery in *Chlamydomonas reinhardtii*. Plant Physiol. **127**: 1394-1398.

Rochaix JR, 2001b, Post-transcriptional control of chloroplast gene expression from RNA to photosynthetic complex. Plant Physiol. **125**: 142-144.

Rockholm DC, Yamamoto HY, 1996, Violaxanthin de-epoxidase: Purification of a 43-kilodalton lumenal protein from lettuce by lipid – affinity precipitation with monogalactosyldiacylglyceride. Plant Physiol. **110**: 697-703.

Rogner M, Boekema EJ, Barber J, 1996, How does photosystem II split water? The structural basis of efficient energy conversion. Trends Biochem. Sci. **21**: 44-49.

Rorat T, Havaux M, Irzykowski W, Cuine S, Becuwe N, Rey P, 2001, PSII-S gene expression, photosynthetic activity and abundance of plastid thioredoxin-related and lipid-associated proteins during chilling stress in *Solanum* species differing in freezing resistance. Physiol. Plant **113**: 72-78.

Rosenquist E, 2001, Light acclimation maintains the redox state of the PSII electron acceptor Q_A within a narrow range over a broad range of light intensities. Photosynth. Res. **70**: 299-310.

Rossel JB, Wilson IW, Pogson BJ (2002) Global changes in gene expression in response to high light in *Arabidopsis*. Plant Physiol. **130**: 1109-1120.

Ruban AV, Horton P, 1994, Spectroscopy of nonphotochemical and photochemical quenching of chlorophyll fluorescence in leaves : evidence for a role of the light harvesting complex of photosystem II in the regulation of energy dissipation. Photosynth. Res. **40**: 181-190.

Ruban AV, Horton P, 1999, The xanthophyll cycle modulates the kinetics of nonphotochemical energy dissipation in isolated light harvesting complexes, intact chloroplasts and leaves of spinach. Plant Physiol. **119**: 531-542.

Ruban AV, Lee PJ, Wentworth M, Young AJ, Horton P, 1999, Determination of the stoichiometry and strength of binding of xanthophylls to the photosystem II light harvesting complexes. J. Biol Chem. **274**: 10458-10465.

Ruban AV, Phillip D, Young AJ, Horton P, 1997, Carotenoid dependent oligomerisation of the major chlorophyll a/b light harvesting complex of photosystem II of plants. Biochemistry **36**: 7855-7859.

Ruban AV, Phillip P, Young AJ, Horton P, 1998, Excited state energy level does not determine the differential effect of violaxanthin and zeaxanthin on chlorophyll fluorescence quenching in the isolated light-harvesting complex of photosystem II. Photochem. Photobiol. **68**: 829-834.

Ruban AV, Wentworth M, Yakushevska AE, Andersson J, Lee PJ, Keegstra W, Dekker JP, Boekema EJ, Jansson S, Horton P, 2003, Plants lacking the main light harvesting complex retain photosystem II macro-organization. Nature **421**: 648-651.

Ruban AV, Young AJ, Horton P, 1993, Induction of nonphotochemical energy dissipation and absorbance changes in leaves : Evidence for changes in the state of the light harvesting system of photosystem II in vivo. Plant Physiol. **102**: 741-750.

Ruf S, Biehler K, Bock R, 2000, A small chloroplast-encoded protein as a novel architectural component of the light-harvesting antennae. J. Cell. Biol. **149**: 369-377.

Ruffle SV, Donnelly D, Blundell TL, Nugent JHA, 1992, A three dimensional model of the photosystem II reaction center *Pisum sativum*. Photosynth. Res. **34**: 287-300

Ruffle SV, Wang J, Johnston HG, Gustafson TL, Hutchison RS, Minagawa J, Crofts Anthony, Sayre RT, 2001, Photosystem II peripheral accessory chlorophyll mutants in *Chlamydomonas reinhardtii*: Biochemical characterization and sensitivity to photo-inhibition. Plant Physiol. **127**: 633-644.

Russel WA, Critchley C, Robinson SA, Frankun LA, Seaton GGR, Chow WS, Anderson JM, Osmond CB, 1995, Photosystem II: regulation and dynamics of the chloroplast D1 protein in *Arabidopsis* leavers during photosynthesis and photoinhibition. Plant Physiol. **107**: 943-952.

Rutherford AW, Paterson DR, Mullet JE, 1981, A light–induced spin-polarized triplet detected by EPR in photosystem II reaction centers. Biochim. Biophys. Aetoc. **635**: 205-14.

Ruuska SA, Badger MR, Andrews TJ, von Caemmerer S, 2000, Photosynthetic electron sinks in transgenic tobacco with reduced amounts of rubisco: Little evidence for significant Mehler reaction. J. Exp. Bot. **51**: 357-368.

Sacksteder CA, Kanazawa A, Jacoby ME, Kramer DM (2000) The proton to electron stoichiometry of steady–state photosynthesis in living plants : A proton–pumping q cycle is continuously engaged. Proc. Natl. Acad. Sci. USA. **97**: 14283-14288.

Sailaja MV, Chandrasekhar D, Rao DN, Das VSR, 1997, Laser-induced chlorophyll fluorescence ratio in certain plants exhibiting leaf heliotropism. Aust. J. Plant Physiol. **24**: 159-164.

Sailaja MV, Das VSR, 1995, Photosystem II acclimation to limiting growth light in fully developed leaves of *Amaranthus hypochondriacus* L, an NAD-ME C_4 plant. Photosynth. Res. **46**: 227-233.

Sailaja MV, Das VSR, 1996a, Leaf solar tracking response exhibits diurnal constancy in photosystem II efficiency. Environ. Exp. Bot. **36**: 431-438.

Sailaja MV, Das VSR, 1996b, Modulation of cytochrome b6/f complex and photosystem I under growth limiting light in *Amaranthus hypochondriacus*, an NAD-ME C_4 plant. Aust. J. Plant Physiol. **23**: 305-309.

Sailaja MV, Das VSR, 2000, Differential photosynthetic acclimation pattern to limiting growth-irradiance in two types of C_4 plants. Photosynthetica **38**: 267-273.

Sakamoto K, Briggs WR, 2002, Cellular and sub-cellular localization of phototropins 1. Plant Cell **14**: 1723-1735.

Santabarbara S, Barbato R, Zucchelli G, Garlaschi FM, Jennings RC, (2001a), The quenching of photosystem II fluorescence does not protect the D1 protein against light induced degradation in thylakoids. FEBS Lett. **505**: 159-162.

Santabarbara S, Bordignon E, Jennings RC, Carbonera D, 2002a, Chlorophyll triplet states associated with photosystem II of thylakoids. Biochemistry **41**: 8184-8194.

Santabarbara S, Cazzalini I, Rivadossi A, Garlaschi FM, Zucchelli G, Jennings CR, 2002b, Photoinhibition in vivo and in vitro involves weakly coupled chlorophyll protein complexes. Photochem. Photobiol. **75**: 613-618.

Santabarbara S, Garlaschi FM, Zucchelli G, Jennings RC, 1999, The effect of excited state population in photosystem II on the photoinhibition–induced changes in chlorophyll fluorescence parameters. Biochim. Biophys. Acta **1409**: 165-170.

Santabarbara S, Neverov KV, Garlaschi FM, Zucchelli G, Jennings RC, 2001b, Involvement of uncoupled antennae chlorophylls in photoinhibition in thylakoids FEBS Lett. **491**: 109-113.

Sapozhnikov DI, Krasnouskaya TA, Maevkaya AN, 1957, Change in the interrelationship of the basic carotenoids of the plastids of green leaves under the action of light. Dokl. Akad. Nauk. USSR **113**: 147-150.

Sazanov LA, Burrows PA, Nixon PJ, 1998, The chloroplast Ndh complex mediates the dark reduction of the plastoquinone pool in response to heat stress in tobacco leaves, FEBS Lett. **429**: 115-118.

Scheller MV, Jensen PE, Haldrup A, Lunde C, Knoetzel J, 2001, Role of subunits in eukaryotic photosystem I. Biochim. Biophys. Acta **1507**: 41-60.

Schindler C, Lichtenthaler HK, 1996, Photosynthetic CO_2- assimilation, chlorophyll fluorescence and zeaxanthin accumulation in field grown maple trees in the course of a sunny and a cloudy day. J. Plant Physiol. **148**: 399-412.

Schnell DJ, 1998, Protein targeting to the thylakoid membrane. Ann. Rev. Plant Physiol. Plant Mol. Biol. **49**: 97-126.

Schreiber U, Bilger W, Hormann M, Neubauer C, 1998, chlorophyll fluorescence as a diagnostic tool: basics and some aspects of practical relevance. In: AS Raghavendra, (ed) Photosynthesis. Cambridge University Press, pp.320-336.

Schreiber V, Schliwa U, Bilger W, 1986, Continuous recording of photochemical and non-photochemical chlorophyll fluorescence quenching with a new type of modulation flourometer. Photosynth. Res. **10**: 51-62.

Schroda M, Vallon O, Wollman FA, Beck CF, 1999, A chloroplast targeted heat shock protein 70 (HSP 70) contributes to the photoprotection and repair of photosystem II during and after photoinhibition. Plant Cell **11**: 1165-1178.

Schubert M, Petersson VA, Haas BJ, Funk C, Schroder WP, Kieselbach T, 2002, Proteome map of the chloroplast lumen of *Arabidopsis thaliana*. J. Biol. Chem **277**: 8354-8365.

Settles AM, Yonetani A, Baron A, Bush DR, Cline K, Martienssen R, 1997, Sec-independent protein translocation by the maize Hcf 106 proteins. Science **278**: 1467-1470.

Sharma J, Panico M, Shipton CA, Nilsson F, Morris HR, Barber J, 1997, Primary structure characterization of the photosystem II D1 D2 subunits. J. Biol. Chem. **272**: 33158-33166.

Sherameti I, Sopory SK, Trebicka A, Pfannschmidt T, Oelmuller R, 2002, Photosynthetic electron transport determines nitrate reductase gene expression and activity in higher plants. J. Biol. Chem. **277**: 46594-46600.

Sherry RA, Galen C, 1998, The mechanism of floral heliotropism in the snow buttercup, *Ranunculus adoneus*. Plant Cell Environ. **21**: 983-993.

Shi LX, Lorkovic ZJ, Oelmuller R, Schroder WP, 2000, The low molecular mass PsbW protein is involved in the stabilization of the dimeric photosystem II complex in *Arabidopsis thaliana*. J. Biol. Chem. **275**: 37945-37950.

Shigeoka S, Ishikawa T, Tamoi M, Miyagawa Y, Takeda T, Yabuta Y, Yoshimura K, 2002, Regulation and function of ascorbate peroxidase isoenzymes. J. Exp. Bot. **53**: 1305-1319.

Shikanai T, Munekago Y, Kimura K, 2002, Regulation of proton-to-electron stiochiometry in photosynthetic electron transport: physiological function in photoprotection. J. Plant Res. **115**: 3-10.

Simpson DV, Knoetzel J, 1996, Light harvesting complexes of plants and algae: Introduction, survey and nomenclature. In: DR Ort and CF Yocum, (eds) Advances in photosynthesis, Vol.4. Kluwer Academic Publishers, Netherlands pp.493-506.

Sinclair J, Park Y, Chow WS, Anderson JM, 1996, Target theory and the photoinactivation of photosystem II. Photosynth. Res. **50**: 33-40.

Smirnoff N, 2001, Biosynthesis of ascorbic acid in plants: A renaissance. Annu. Rev. Plant Physiol. Plant Mol. Biol. **52**: 437-467.

Smirnoff N, 2002, Antioxidants and reactive oxygen species in plants. J. Exp. Bot. **53**: p.IV (preface).

Smirnoff N, 2000, Ascorbate biosynthesis and function in photoprotection. Phil.Trans. R.Soc. Lond. B. **355**: 1455-1464.

Smith H, 1984, Plants that track the sun. Nature **308**: 774.

Smith TA, Kohorn BD, 1994, Mutations in a signal sequence for the thylakoid membrane identify multiple prolesin transport pathways and nuclear suppressors. J. Cell. Biol. **126**: 365-374.

Snyders S, Kohorn B, 2001, Disruption of thylakoid – associated kinase 1 leads to ateration of light havesting in *Arabidopsis*. J. Biol. Chem. **276**: 32169-32176.

Somerville C, Somerville S, 1999, Plant functional genomics Science. **285**: 380-383.

Sonoike K, 1995, Selective photoinhibition of photosystem I in isolated thylakoid membranes from cucumber and spinach. Plant Cell Physiol. **36**: 825-830.

Sonoike K, 1996, Degradation of PsaB gene product, the reaction center subunit of photosystem I: possible involvement of active oxygen speices. Plant Sci. **115**: 157-164.

Sonoike K, Kamo M, Hihara Y, Hiyama T, Enami I, 1997, The mechanism of the degradation of PsaB gene product, one of the photosynthetic reaction center subunits of photosystem I, report on photoinhibition. Photosynth. Res. **53**: 55-63.

Sonoike K, Terashima I, 1994, Mechanism of the photosystem I photoinhibition in leaves of *Cucumis sativus* L. Planta **194**: 287-293.

Spetea C, Hundae T, Lohmann F, Andersson B, 1999, GTP bound to chloroplast thylakoid membrane is required for light induced multienzyme degradation of the photosystem II D1 protein. Proc. Natl. Acad. Sci. USA **96**: 6547-6552.

Srivastava A, Zeiger E, 1995, Guard cell zeaxanthin tracks photosynthetically active radiation and stomatal apertures in V*icia faba* leaves. Plant Cell Environ. **18**: 813-817.

Streb P, Feierabend J, 1999, Significance of antioxidants and electron sinks for the cold – hardening – induced resistance of winter rye leaves to photo-oxidative stress. Plant Cell Environ. **22**: 1225-1237.

Streb P, Shang W, Feierabend J, Bligny R, 1998, Divergent strategies of photophotection in high-mountain plants. Planta **207**: 313-324.

Styring S, Jegerschold C, 1994, Light–induced reactions impairing electron transfer through photosystem II. In: NR Baker, JR Bowyer (eds) Photoinhibition of Photosynthesis Bios Scientific Publishers Oxford, pp.51-73.

Sugiura M, Hirose T, Sugila M, 1998, Evolution and mechanism of translation in chloroplasts. Annu. Rev. Gent. **32**: 437-459.

Sun WH, Verhoeven AS, Bugos RC, Yamamoto HY, 2001, Suppression of zeaxanthin formation does not reduce photosynthesis and growth of transgenic tobacco under field conditions. Photosynth. Res. **67**: 41-50.

Svab Z, Maliga P, 1993, High-frequency plastid transformation in tobacco by selection for a chimeric A gene. Proc. Natl. Acad. Sci. USA **90**: 913-917.

Swiatek M, Kuras R, Sokolenko A, Higgs D, Olive J, Cinque G, Muller B, Eichacker LA, Stern DB, Bassi R, Herrmann RG, Wollman FA, 2001, The chloroplast gene ycf9 encodes a photosystem II (PSII) core subunit PsbZ that participates in PSII supramolecular architecture. Plant Cell **13**: 1347-1367.

Tambussi EA, Casadesces J, Munne Bosun S, Araus JL, 2002, Photoprotection in water-stressed plants of durum wheat (*Triticum turgidum* var durum). Changes in chlorophyll fluorescence, spectral signature and photosynthetic pigments. Funct. Plant Biol. **29**: 36-44.

Tanaka R, Oster U, Kruse E, Rudigar W, Grimm B, 1999, Reduced activity of geranyl reductase leads to loss of chlorophyll and tocopherol to partically geranylgeranylated chlorophyll in transgenic tobacco plants expressing antisense RNA for geranylgeranyl reductase. Plant Physiol. **120**: 695-704.

Telfer A, Barber J, 1994, Elucidating the molecular mechanisms of photoinhibition by studying isolated PS II reaction centers. *In:* NR Baker, Bowyer JR, (eds) Photoinhibition of Photosynthesis. From

Molecular Mechanisms to Field. Bios Scientific Publishers Oxford, pp.22-49.

Telfer A, Bishop SM, Phillips D, Barber J, 1994, Isolated photosynthetic reaction catre of photosystem II as a sansitizer for the formation of singlet oxygen. Detection and quantum yield determination using a chemical trapping technique. J. Biol. Chem. **269**: 13244-13253.

Teramoto H, Nakamoli A, Minayawa J, Ono TA, 2002, Light-intensity dependent expression of Lhc gene family encoding light-harvesting chlorophyll a/b proteins of photosystem II in *Chlamydomonas reinhardtii*. Plant Physiol. **130**: 325-333.

Terashima I, Funayama S, Sonoike K, 1994, The site of photoinhibition in leaves of *Cucumis sativus* L. at low temperature is photosystem I, not photosystem II. Planta **193**: 300-306.

Terashima I, Noguchi K, Itoh – Xlenote T, Park YM, Kubo A, Tanaka K, 1998, The cause of PS I photoinhibition at low temperatures in leaves of *Cucumis satibus*, a chilling-sensitive plant. Physiol Plant **103**: 295-303.

Thiele A, Krause GH (1994) Xanthophyll cycle and thermal energy dissipation in photosystem II : Relationship between zeaxanthin formation, energy dependent fluorescence quenching and photoinhibition. J. Plant Physiol. **144**: 324-332.

Thiele A, Schirwitz K, Winter K, Krause GH, 1996, Increased xanthophyll cycle activity and reduced D1 protein inactivation related to photoinhibition in two plant systems acclimated to excess light. Plant Sci. **115**: 237-250.

Thornber JP, Cogdell RJ, Chitnis P, Morshige DT, Peter GF, Gomez SM, Anandan S, Preiss S, Dreyfuss BW, Lee A, Takeuchi T, Kerfeld C, 1994, Antennae pigment – protein complexes of higher plants and purple bacteria. Adv. Mol. Cell. Biol. **10**: 55-118.

Tjus SE, Moller BL, Sheller HV, 1999, Photoinhibition of photosystem I damages both reaction center proteins PS I–A and PS I–B and acceptor side located small photosystem I polypeptide. Photosynth. Res. **60**: 75-86.

Tjus SE, Scheller HV, Andersson B, Moller BL, 2001, Active oxygen produced during selective excitation of photosystem 1, is damaging not only to photosystem I but also to photosystem II. Plant Physiol. **125**: 2007-2015.

Tracewell CA, Vrettos JS, Bautista JA, Frank HA, Brudvig GW, 2001, Carotenoid photo-oxidation in photosystem II. Arch. Biochem. Biophys. **385**: 61-69.

Trebst A, 1986, The topology of the plastoquinone and herbicide binding peptides of photosystem II in the thylakoid membranes. Z, Naturforsch. **41C**: 240-245.

Tremoliers A, Dainese P, Bassi R, 1994, Heterogeneous lipid distribution among chlorophyll–binding proteins of photosystem II in maize mesophyll chloroplasts. Eur. J. Biochem. **221**: 721-730.

Tsiotis G, Psylinakis M, Woplensinger B, Lustig A, Engel A, Ghanotakis D, 1999, Investigation of the structure of spinach photosystem II reaction center complex. Eur. J. Biochem. **259**: 320-324.

Tyystjarvi E, Kettunen R, Aro EM, 1994, The rate constant of photo inhibition in vitro is independent of the antennae size of photosystem II but depends on temperature. Biochim. Biophys. Acta **1186**: 177-185.

Tyystjarvi E, King N, Hakala M, Aro EM, 1999a, Artificial quenchers of chlorophyll fluorescence do not protect against photoinhibition. J. Photochem. Photobiol. B. **48**: 142-147.

Tyystjarvi E, Riikonen M, Arisi AM, Kettunen R, Jouanin L, Foyer CH, 1999b, Photoinhibition of photosystem II in tobacco plants over expressing glutathione reductase and poplars overexpressing superoxide dismutase. Physiol Plant. **105**: 409-416.

van Kooten O, Snel JFH, 1990, The use of chlorophyll fluorescence nomenclature in plant stress physiology. Photosynth. Res. **25**: 147-150.

van Wijk KJ, 2000, Proteomics of the chloroplast: experimentation and prediction. Trends Plant Sci. **5**: 420-425.

van Wijk KJ, Robol – Boza M. Kettunen R, Andersson B, Aro EM, 1997, Synthesis and assembly of the D1 protein into photosystem II : Processing of C-terminus and identification of the initial assembly partners and complexes during photosynthesis II repair. Biochemistry **36**: 6178-6186.

Varotto C, Pesaresi P, Jahns P, Lebnict A, Tizzano M, Schiavon F, Salamini F, Leister D, 2002, Single and double knockouts of the genes for photosystem I subunits G K, and H of *Arabidopsis*. Effects on photosystem I composition, photosynthetic electron flow, and state transititions. Plant Physiol. **129**: 616-624.

Vasilev S, Orth P, Zouni A, Owens TG, Bruce D, 2001, Excited state dynamics in photosystem II: insights from the x-ray crystal structure. Proc. Natl. Acad. Sci. USA **98**: 8602-8607.

Vass I, Styring S, Hundall T, Koirnniemi A, Aro EM, Andersson B, (1992) The reversible and irreversible intermediates during photoinhibition of photosystem II – stable reduced Q species promote chlorophyll triplet formation. Proc. Natl. Acad. Sci. USA **89**: 1408-1412.

Veeranjaneyulu K, Charland M, Leblanc RM, 1998, High irradiance stress and photochemical activities of photosystems 1 and 2 in vivo. Photosynthetica. **35**: 177-190.

Verhoeven AS, Adams III WW, Demming–Adams B, Croce R, Bassi R, 1999, Xanthophyll cycle pigment localization and dynamics during exposure to low temperatures and light stress in *Vinca major*. Plant Physiol. **120**: 727-737.

Verhoeven AS, Bugos RC, Yamamoto HY, 2001, Transgenic tobacco with suppressed zeaxanthin formation is susceptible to stress-induced photoinhibition. Photosynth, Res. **67**: 27-39.

Von Caemmerer S, Farguhar GD, 1984, Effects of partial defoliatio, changes of irradiance during growth, short term water stress and growth at enhanced P (CO_2) on the photosynthetic capacity of leaves of *Phascolus uelgaris*, L. Planta **160**: 320-329.

Von Caemmerer S, Farquhar GD, 1984, Effects of partial defoliation, changes of irradiance during growth, short-term water stress and growth at enhanced $p(CO_2)$ on the photosynthetic capacity of leaves of *Phaseolus vulgaris* L. Planta **160**: 320-329.

Vranova E, Inze D, Brellsegem FV, 2002, Signal transduction during oxidative stress. J. Exp. Bot. **53**: 1227-1236.

Walters RG, Horton P, 1994, Acclimation of *Arabidopsis thaliana* to the light environment: Changes in composition of the photosynthetic apparatus. Planta **195**: 248-256.

Walters RG, Horton P, 1995a, Acclimation of *Arabidopsis thaliatha* to the light environment: Changes in photosynthetic function. Planta **197**: 306-312.

Walters RG, Rogers JJM, Shephard F, Horton P, 1999, Acclimation of *Arabidopsis thaliana* to the light environment: the role of photoreceptors. Planta **209**: 517-527.

Weis E, Berry JA, 1987, Quantum efficiency of photosystem II in relation to 'energy' – dependent quenching of chlorophyll fluorescence. Biochim. Biophys. Acta **894**: 198-208.

Wentworth M, Ruban AV, Horton P, 2000, Chlorophyll fluorescence quenching in isolated light harvesting complexes induced by zeaxanthin. FEBS Lett. **471**: 71-74.

Werner C, Correia O, Beyschlag, 2002, Characteristic patterns of chronic and dynamic photoinhibition of different functional groups in a mediterranean ecosystem. Funct. Plant Biol. **29**: 999-1011.

Weston E, Thorogood K, Vinti G, Lopez-Juez E, 2000, Light quantity controls leaf-cell and chloroplast development in *Arabidoposis thaliana* wild type and blue light perception mutants. Planta **211**: 807-815.

Wheeler GL, Jones MA, Smirnoff N, 1998, The biosynthetic pathway of vitamin C in higher plants. Nature **393**: 365-369.

Wiese C, Shi L-B, Heber U, 1998, Oxygen reduction in the Mehler reaction is insufficient to protect photosystems I and II of leaves against photo-inactivation. Physiol Plant. **102**: 437-446.

Wild A, Hopfner M, Ruhle W, Richter M, 1986, Changes in the stoichiometry of Photosystem II components as an adaptive response to high light and low light conditions during growth. Z. Naturforsch. C. **41**: 597-603.

Wilson KE, Huner NPA, 2000, The role of growth rate, redox state of the plastoquinone pool and the trans-thylakoid pH in photoacclimation of *Chlorella vulgaris* to growth irradiance and temperature. Planta **212**: 93-102.

Wingler A, Lea PJ, Quick PW, Leegood RC, 2000, Photorespiration: metabolic pathways and their role in stress protection. Phil. Trans. R. Soc. Lond. B. **355**: 1517-1529.

Xiang C, Werner BL, Christensen EM, Oliver DJ, 2001, The biological functions of glutathione revisited in *Arabidopsis* transgenic plants with altered glutathione leaves. Plant Physiol. **126**: 564-574.

Xu CC, Li L, Kuang T, 2000, The inhibited xanthophyll cycle is responsible for the increase in sensitivity to low temperature photo inhibition in rice leaves fed with glutathione. Photosynth. Res. **65**: 107-114.

Yamamoto HY, 1979, Biochemistry of the violaxanthin cycle in higher plants. Pure Appl. Chem. **51**: 639-648.

Yamamoto HY, Bassi R, 1996, Carotenoids localization and function in oxygenic photosynthesis. In: DR Ort, CF Yocum, (eds) Advances in Photosynthesis. Vol.4. Kluwer Academic Publishers, Dordrecht, pp.539-563.

Yamamoto HY, Bugos RC, Hieber AD, 1999, Biochemistry and molecular biology of the xanthophyll cycle. In: HA Frank, AJ Young, G Britton, RJ Cogdell (eds) Advances in Photosynthesis. Vol.8. Kluwer Academic Publishers, Dordrecht. pp.293-303.

Yamamoto HY, Kamite L, 1972, The effects of dithiothreitol on violaxanthin de-epoxidation and absorbance changes in the 500 nm region. Biochim. Biophys. Acta **267**: 538-543.

Yamamoto HY, Nakayama TOM, Chichester CO, 1962, Studies on the light and dark interconversions of leaf xanthophylls. Arch. Biochem. Biophys. **97**: 168-173.

Yamamoto Y, 2001, Quality control of photosystem II. Plant Cell Physiol. **42**: 121-128.

Yamamoto Y, Ishikawa Y, Nakatani E, Yamada M, Zhang H, Wydrzynski T, 1998, Role of an extrinsic 33 kilodalton protein of photosystem II to the turn over of the reaction centre – binding protein D1 during photoinhibition. Biochemistry. **37**: 1565-1574.

Yamasato A, Kamada T, Satoh K, 2002, Random mutagenesis targeted to the psb A II gene of *Syneehocystis* sp PCC 6803 to identify functionally important residues in the D1 protein of the photosystem II reaction center. Plant Cell Physiol. **43**: 540-548.

Yamashita N, Ishida A, Kushima H, Tanaka N, 2000, Acclimation to sudden increase in light favouring an invasive over native/trees in subtropical islands, Japan. Occologia **125**: 412-419.

Yamomoto Y, Akasaka T, 1995, Degradation of antennae chlorophyll II – binding protein CP 43 during photoinhibition of photosystem II. Biochemistry **43**: 9038-9045.

Yang C, Harn R, Paulsen H, 2003, The light – harvesting chlorophyll a/b complex can be reconstituted in vitro from its completely unfolded apoprotein. Biochemistry **42**: 4527-4533.

Yang DH, Andersson B, Aro EM, Ohad I, 2001, The redox state of the plastoquinone pool controls the level of the light-harvesting chlorophyll a/b binding protein complex II (LHCII) during photoacclimation. Photosynth. Res. **68**: 163-174.

Yang DH, Paulsen H, Andersson B, 2000, The N terminal domain of the light-harvesting chlorophyll a/b binding protein complex (LHCII) is essential for its acclimative proteolysts. FEBS Lett. **466**: 385-388.

Yang DH, Webster J, Adam Z, Lindahl M, Andersson B, 1998, Induction of acclimative proteolysis of the light-harvesting chlorophyll a/b protein of photosystem II in response to elevated light intensities. Plant Physiol. **118**: 827-834.

Yin HC (1938) Diaphototropic movement of the leaves of *Malva negleeta*. Am. J. Bot. **25**: 1-6.

Yin ZH, Johnson GN, 2000, Photosynthetic acclimation of higher plants to growth in fluctuating light environments. Photosynth. Res. **63**: 97-107.

Yoder LM, Cole AG, Sension RJ, 2002, Structure and function in the isolated reaction center complex of photosystem II: energy and charge transfer dynamics and mechanism Photosynth. Res. **72**: 147-158.

Yokthongwattana K, Chrost B, Behrman S, Casper-Lindley C, Melis A, 2001, Photosystem II damage and repair cycle in the green algae *Dunaliella salina:* Involvement of a chloroplast localized HSP70. Plant Cell Physiol. **42**: 1389-1397.

Young AJ, Frank HA, 1996, Energy transfer reactions involving carotenoids: quenching of chlorophyll fluorscence. J. Photochem. Photobiol. **36**: 3-15.

Young AJ, Phillip D, Frank HA, Ruban AV, Horton P, 1997, The xanthophyll cycle and carotenoid mediated dissipation of excess excitation energy in photosynthesis. Pure Appl. Chem. **69**: 2125-2130.

Yu F, Berg VS, 1994, Control of paraheliotropism in two *Phaseolus* species. Plant Physiol **106**: 1567-1573.

Zhang H, Goodman HM, Jansson S, 1997, Antisense inhibition of the photosystem I antennae protein Lhca 4 in *Arabidopsis thaliana*. Plant Physiol. **115**: 1525-1531.

Zhang L, Aro EM, 2002, Synthesis, membrane insertion and assembly of the chloroplast encoded D1 protein into photosystem II FEBS Lett. **512**: 13-18.

Zhang L, Paakkarenen V, Wijk KJ, Aro EM, 1999, Co-translational assembly of the D1 protein into photosystem II. J. Biol. Chem. **274**: 16062-16067.

Zhang L, Paakkarinen V, Soursa M, Aro EM (2001) A Sey homologue is involved in chloroplast – encoded D1 protein biogenesis. J. Biol. Chem. **276**: 37809-37814.

Zhang L, Paakkarinen V, Wizk KJ, Aro EM, 2000, Biogensis of the chloroplast encoded D1 protein. Regulation of translation elongation, insertion and assembly into photosystem II. Plant Cell **12**: 1769-1781.

Zheng B, Halperin T, Hruskova-Heidings feldava O, Adam Z, Clarke AK, 2002, Characterization of chloroplast Clp proteins in *Arabidopsis*. Localization, tissue specificity and stress responses. Physiol Plant. **114**: 92-101-2002.

Zolla L, Rinalducci S, Timperio AM, Huber CG, 2002, Proteomics of light harvesting proteins in different plants species. Analysis and comparison by liquid chromatography–electrospray ionization mass spectrometry. Photosystem I. Plant Physiol. **130**: 1938-1950.

Zolla L, Timperio AM, Walcher W, and Huber CG, 2003, Proteomilcs of light-harvesting proteins in different plant species. Analysis and comparison by liquid chromatography-electrospray ionization mass spectrometry. Photosystem II. Plant Physiol. **131**: 198-214.

Zouni A, Witt HT, Karn J, Fromme P, Krauss N, Saenger W, Orth P, 2001, Crystal structure of photosystem II from *Synechococcus elongatus* at 3.8 Å resolution. Nature **409**: 739-743.

Author Index

A

Abdallah F 45, 122
Adam Z 37, 122
Adams III WW 3, 27, 47, 48, 49, 50, 51, 52
Adir N 26, 35, 40, 122
Agati G 90, 122
Akasaka T 39
Akerlund HE 64, 65
Albertsson PA 2, 122
Alia 113, 122
Allen JF 114, 122
Alves PL 1, 26, 47, 84, 122
Ananyev GM 17, 122
Anderson JM 28, 34, 48, 49, 72, 98, 122, 123
Andersson B 26, 27, 30, 31, 32, 34, 35, 36, 123
Andersson J 52, 109, 110, 123
Apel K 72
Aro EM 27, 33, 34, 36, 40, 42, 43, 44, 124
Asada K 49, 71, 72, 74, 77, 124
Augusti A 73

B

Babcock GT 2
Badger M 82, 124
Baena – Gonzalez E 39, 124
Bailey S 37, 94, 96, 101, 125
Baker NRI 1, 34, 47, 48, 53, 78, 79, 119, 125
Barbato R 39, 125
Barber J 13, 15, 18, 19, 35, 36, 38, 125
Barkan A 117, 125
Baroli I 69, 115, 125
Barry BA 16, 125

Barter LMC 12, 126
Barth C 80, 81, 89, 126
Bassi R 4, 7, 8, 58, 65, 125
Batschauer A 93, 126
Bell CJ 1, 126
Bennoun P 81, 126
Berg VS 84, 85, 86
Biehler K 79, 126
Bilger W 52, 68, 126
Bjorkman O 48, 52, 84, 96, 126
Blubaugh DJ 33, 126
Boardman NK 94, 95, 126
Boekema EJ 17, 126
Boutry M 93
Bratt CE 64, 127
Brettel K 21, 127
Brezina V 27
Bricker JM 4, 5, 17, 18, 127
Briggs WR 93, 127
Britton G 67, 127
Brugnoli E 48, 127
Buch K 64, 127
Bugos RC 66, 127
Bukhov NG 56, 57, 127
Burkey KO 96, 98
Burrows PA 81, 127

C

Caffari S 55, 128
Cai SQ 35, 128
Canovas PM 38, 128
Carol P 81, 128
Casano LM 81, 128
Chappel EW 90, 128
Chassin Y 38, 128
Chitnis PR 19, 128
Chow WS 59, 95, 128

Christie JM 93, 129
Clarke JE 80, 82, 129
Cleland RC 51, 129
Cline K 40, 42
Coley PD 98
Cornic G 80, 129
Creighton AM 43, 129
Critchley C 27, 129
Croce R 6, 129
Crofts AR 52, 59, 129
Cronlund SL 86, 93, 129

D

Dainese P 4, 7
Dalbey RE 42, 129
Danon A 42, 129
Das VS Rama 48, 85, 87, 91, 96, 97, 98, 99, 100, 101, 120
Davaud A 26
Davis EC 98
De la Torre WR 96, 98, 129
De Las Rivas J 35, 37, 130
Debus RJ 16, 130
Decoster B 55, 67, 130
Della Penna D 72
Demmig-Adams B 3, 12, 27, 46, 47, 49, 51, 52, 60, 61, 66, 67, 68, 83, 89, 95, 130, 131
Dent RM 115, 117, 131
Depka B 45, 131
Dietz KJ 72
Diner BA 12, 14, 131
Dominici P 6, 11, 120, 131
Donahue R 85, 131
Dreyfuss BW 2, 6, 10
Durnford DG 8, 10

E

Ebbert V 53, 131
Eckardt NA 107, 131
Eckert HJ 33, 131
Ehleringer J 85, 86, 87, 131
Elrad D 59, 120, 131
Endo T 81, 132
Escoubas JM 103, 133
Eshagi S 107, 132

Eskling M 62, 64, 65, 66, 68, 69, 132
Evans JR 94, 96, 132

F

Falkowiski PG 94, 132
Faller P 17, 19
Farah J 117
Farber A 66, 132
Farquhar GD 98
Ferrar PJ 98
Field TS 82, 132
Fierabend J 79
Finazzi G 29, 82, 133
Fischer M 82, 133
Flachman R 8, 110, 133
Fock H 79
Formaggio E 2, 6, 11, 133
Forsberg J 114
Forseth IN 48, 85, 86, 89, 93, 133
Forster B 57, 133
Fotinou C 17, 29, 44, 133
Foyer C 49, 71, 74, 75, 76, 77, 133, 134
Frank HA 27, 54, 55, 67, 68, 134
Frankel LK 4, 5, 18
Frechilla S 70, 134
Fricker MD 72
Fryer MJ 73, 79, 134
Fujita Y 95, 104, 134
Funk C 9, 10, 134

G

Galan C 85
Ganetag U 109, 110, 134
Garab G 53, 134
Gareia-Plazaola JI 71, 75, 120, 135
Genty B 79, 95, 135
Gerst U 80, 135
Ghanotakis DF 5, 17, 29, 44, 135
Gilmore AM 49, 59, 61, 62, 65, 66, 67, 69, 134, 135
Gnatt E 94
Golbeck JH 19, 23, 25, 136
Goldschmidt-clermont M 117
Gonzalez EB 35, 136
Gonzalez-Rodriguez AM 73, 136
Govindjee 48, 49, 52, 54, 57, 58, 120, 136

Grace SC 73, 79, 81, 136
Grasses T 59, 70, 112, 113, 136
Gray GR 27, 136
Green BR 102, 136
Green BR 8, 10, 136
Greer DH 34, 137
Grotz B 62, 137
Grusak MA 72, 137
Gullner G 72, 137

H

Hager A 62, 137
Hak R 90, 137
Haldrup A 83, 108, 137
Hanba YT 98, 137
Hankamer B 13, 14, 15, 16, 29, 35, 36, 137, 138
Harbinson J 72, 79, 95
Harrer R 19, 138
Hartel H 57, 61, 138
Haussuhl K 37, 38, 138
Havaux M 3, 11, 26, 52, 56, 69, 70, 138
Havir EA 52, 56, 74, 138
Havir EA 62, 138
He WZ 17, 19, 20, 138
Heber U 71, 79, 80, 82, 139
Henmi T 18, 39, 139
Hieber AD 112, 139
Hihara Y 26, 103, 139
Hihara Y 45, 46, 139
Hihara Y 94, 96, 104
Hillier W 2, 139
Horton P 2, 3, 5, 6, 8, 47, 49, 51, 52, 53, 57, 58, 59, 65, 69, 94, 101, 114, 140
Hirata M 84, 139
Hirose T 42, 139
Hirschberg J 12, 139
Houben E 43, 140
Howitt CA 80, 140
Huber CG 7, 140
Huner NPA 3, 103, 104, 105, 140
Hurry V 26, 55, 69, 140

I

Ihalainen JA 108, 140
Ishikawa Y 39, 140

Ivanov B 75, 141

J

Jackowski G 109, 141
Jacob B 71, 141
Jahns P 34, 59, 61, 66, 141
Jansson S 2, 5, 6, 9, 10, 109, 141
Jarvis PG 104, 141
Jegerdschold C 30, 33, 142
Jensen PK 20, 108, 142
Jeong WJ 113, 142
Jin E 71, 142
Joet T 80, 142
Johnson GN 53, 80, 82, 101, 102, 103, 121
Jordon P 14, 19, 21, 22, 25
Jurik 98, 142

K

Kagawa T 48, 142
Kamite L 62
Kamiya N 13, 18, 142
Kaneko T 115, 118, 142
Kao W 48, 85, 86, 89, 92, 142
Karpinski S 3, 47, 48, 71, 75, 79, 105, 143
Kasahara M 49, 143
Kato MC 26, 33, 34, 119
Ke B 1, 29, 30, 143
Keegstra K 40, 42, 143
Kettunen R 33, 143
Khorobrykh SA 75, 144
Kim JH 41, 144
Kim JH 95, 103
Kimura M 115, 144
Kitao M 27, 47, 144
Kloppstech K 52
Knoetzel J 2, 7, 10, 144
Kogata N 42, 43
Kohorn B 2, 12, 42, 83, 114
Kok B 26, 144
Koller D 48, 84, 85, 87, 92, 120, 144
Kozaki A 80
Krause GH 27, 33, 34, 49, 53, 59, 69, 80, 81, 89, 144
Krauss N 14, 145
Krieger A 56, 145
Krol M 10, 145

Kuhlbrandt W 6, 8, 10, 13, 14, 15, 145
Kulheim C 74, 145
Kurar TA 98, 144

L

Lagoutte B 24, 145
Lancaster CRD 21, 145
Laroche J 94
Lawson T 52, 144
Lazar D 52, 146
Lee HY 34, 146
Leong SP 96, 97, 146
Levy I 94, 146
Li XP 10, 53, 54, 60, 69, 120, 146
Lichtenthaler HK 70, 90, 146
Lindahl M 36, 37, 38, 146
Lloyd JC 106
Logan A 73, 77, 146, 147
Lokstein 0, 66, 147
Long SP 1, 26, 47, 48, 51, 84, 86, 147

M

Maenpaa P 51, 147
Malkin R 17, 19, 20
Malliga P 117
Mant A 42
Marin E 70, 147
Marshall HL 35, 147
Masuda T 103, 147
Maxwell K 102
Maxwell K 53, 147
Mayfield SP 41
Medrano H 119
Mehler AH 77, 147
Meir P 104, 147
Melis A 2, 11, 12, 28, 32, 33, 51, 52, 71, 95, 103, 148
Meyer WS 72, 84, 148
Michelet B 93, 148
Miyake E 81, 119, 148
Mo F 14, 148
Morais F 43, 148
Morishige DT 2, 6, 10, 148
Muller P 47, 51, 52, 149
Mullineaux P 3, 47, 71, 79, 81, 149
Mulo P 30, 149

Munekage Y 53, 149
Murakami A 104, 149
Murchie EH 28, 94, 149
Murgia I 72, 149

N

Nanba O 17, 29, 149
Naver H 116, 149
Neubauer C 62, 149
Nield J 12, 15, 150
Nilson R 41, 43, 150
Nishikami H 72, 150
Nixon PJ 81, 150
Niyogi KK 2, 12, 47, 51, 56, 61, 69, 70, 77, 81, 84, 114, 115, 119, 150
Noctor G 58, 71, 72, 75, 77, 150

O

Obokata J 116, 151
Ogren E 28, 151
Oguchi R 84, 98, 121, 151
Ohad I 26, 33, 151
Ohnishi N 41, 151
Okamura M 119
Oosterhuis DM 87, 151
Oquist G 34, 151
Ort DR 3, 4, 47, 48, 49, 51, 53, 78, 79, 119, 151
Ortega JM 39, 151
Osmond CB 27, 50, 79, 81, 94, 95, 151
Oswald O 114, 152
Ott T 79, 152
Owens TG 2, 4, 27, 53, 54, 67, 152

P

Palatnik J 79, 152
Palmer JM 71, 152
Park YII 35. 48, 51, 152
Paulsen H 2, 4, 5, 6, 8, 152
Peltier G 116, 152
Pesaresi P 106, 115, 117, 152
Peterson RB 52, 56, 74, 153
Pfannschimdt T 12, 116, 153
Pfundel EE 68, 153
Phillip Y 66, 153

Grace SC 73, 79, 81, 136
Grasses T 59, 70, 112, 113, 136
Gray GR 27, 136
Green BR 102, 136
Green BR 8, 10, 136
Greer DH 34, 137
Grotz B 62, 137
Grusak MA 72, 137
Gullner G 72, 137

H

Hager A 62, 137
Hak R 90, 137
Haldrup A 83, 108, 137
Hanba YT 98, 137
Hankamer B 13, 14, 15, 16, 29, 35, 36, 137, 138
Harbinson J 72, 79, 95
Harrer R 19, 138
Hartel H 57, 61, 138
Haussuhl K 37, 38, 138
Havaux M 3, 11, 26, 52, 56, 69, 70, 138
Havir EA 52, 56, 74, 138
Havir EA 62, 138
He WZ 17, 19, 20, 138
Heber U 71, 79, 80, 82, 139
Henmi T 18, 39, 139
Hieber AD 112, 139
Hihara Y 26, 103, 139
Hihara Y 45, 46, 139
Hihara Y 94, 96, 104
Hillier W 2, 139
Horton P 2, 3, 5, 6, 8, 47, 49, 51, 52, 53, 57, 58, 59, 65, 69, 94, 101, 114, 140
Hirata M 84, 139
Hirose T 42, 139
Hirschberg J 12, 139
Houben E 43, 140
Howitt CA 80, 140
Huber CG 7, 140
Huner NPA 3, 103, 104, 105, 140
Hurry V 26, 55, 69, 140

I

Ihalainen JA 108, 140
Ishikawa Y 39, 140

Ivanov B 75, 141

J

Jackowski G 109, 141
Jacob B 71, 141
Jahns P 34, 59, 61, 66, 141
Jansson S 2, 5, 6, 9, 10, 109, 141
Jarvis PG 104, 141
Jegerdschold C 30, 33, 142
Jensen PK 20, 108, 142
Jeong WJ 113, 142
Jin E 71, 142
Joet T 80, 142
Johnson GN 53, 80, 82, 101, 102, 103, 121
Jordon P 14, 19, 21, 22, 25
Jurik 98, 142

K

Kagawa T 48, 142
Kamite L 62
Kamiya N 13, 18, 142
Kaneko T 115, 118, 142
Kao W 48, 85, 86, 89, 92, 142
Karpinski S 3, 47, 48, 71, 75, 79, 105, 143
Kasahara M 49, 143
Kato MC 26, 33, 34, 119
Ke B 1, 29, 30, 143
Keegstra K 40, 42, 143
Kettunen R 33, 143
Khorobrykh SA 75, 144
Kim JH 41, 144
Kim JH 95, 103
Kimura M 115, 144
Kitao M 27, 47, 144
Kloppstech K 52
Knoetzel J 2, 7, 10, 144
Kogata N 42, 43
Kohorn B 2, 12, 42, 83, 114
Kok B 26, 144
Koller D 48, 84, 85, 87, 92, 120, 144
Kozaki A 80
Krause GH 27, 33, 34, 49, 53, 59, 69, 80, 81, 89, 144
Krauss N 14, 145
Krieger A 56, 145
Krol M 10, 145

Grace SC 73, 79, 81, 136
Grasses T 59, 70, 112, 113, 136
Gray GR 27, 136
Green BR 102, 136
Green BR 8, 10, 136
Greer DH 34, 137
Grotz B 62, 137
Grusak MA 72, 137
Gullner G 72, 137

H

Hager A 62, 137
Hak R 90, 137
Haldrup A 83, 108, 137
Hanba YT 98, 137
Hankamer B 13, 14, 15, 16, 29, 35, 36, 137, 138
Harbinson J 72, 79, 95
Harrer R 19, 138
Hartel H 57, 61, 138
Haussuhl K 37, 38, 138
Havaux M 3, 11, 26, 52, 56, 69, 70, 138
Havir EA 52, 56, 74, 138
Havir EA 62, 138
He WZ 17, 19, 20, 138
Heber U 71, 79, 80, 82, 139
Henmi T 18, 39, 139
Hieber AD 112, 139
Hihara Y 26, 103, 139
Hihara Y 45, 46, 139
Hihara Y 94, 96, 104
Hillier W 2, 139
Horton P 2, 3, 5, 6, 8, 47, 49, 51, 52, 53, 57, 58, 59, 65, 69, 94, 101, 114, 140
Hirata M 84, 139
Hirose T 42, 139
Hirschberg J 12, 139
Houben E 43, 140
Howitt CA 80, 140
Huber CG 7, 140
Huner NPA 3, 103, 104, 105, 140
Hurry V 26, 55, 69, 140

I

Ihalainen JA 108, 140
Ishikawa Y 39, 140

Ivanov B 75, 141

J

Jackowski G 109, 141
Jacob B 71, 141
Jahns P 34, 59, 61, 66, 141
Jansson S 2, 5, 6, 9, 10, 109, 141
Jarvis PG 104, 141
Jegerdschold C 30, 33, 142
Jensen PK 20, 108, 142
Jeong WJ 113, 142
Jin E 71, 142
Joet T 80, 142
Johnson GN 53, 80, 82, 101, 102, 103, 121
Jordon P 14, 19, 21, 22, 25
Jurik 98, 142

K

Kagawa T 48, 142
Kamite L 62
Kamiya N 13, 18, 142
Kaneko T 115, 118, 142
Kao W 48, 85, 86, 89, 92, 142
Karpinski S 3, 47, 48, 71, 75, 79, 105, 143
Kasahara M 49, 143
Kato MC 26, 33, 34, 119
Ke B 1, 29, 30, 143
Keegstra K 40, 42, 143
Kettunen R 33, 143
Khorobrykh SA 75, 144
Kim JH 41, 144
Kim JH 95, 103
Kimura M 115, 144
Kitao M 27, 47, 144
Kloppstech K 52
Knoetzel J 2, 7, 10, 144
Kogata N 42, 43
Kohorn B 2, 12, 42, 83, 114
Kok B 26, 144
Koller D 48, 84, 85, 87, 92, 120, 144
Kozaki A 80
Krause GH 27, 33, 34, 49, 53, 59, 69, 80, 81, 89, 144
Krauss N 14, 145
Krieger A 56, 145
Krol M 10, 145

Grace SC 73, 79, 81, 136
Grasses T 59, 70, 112, 113, 136
Gray GR 27, 136
Green BR 102, 136
Green BR 8, 10, 136
Greer DH 34, 137
Grotz B 62, 137
Grusak MA 72, 137
Gullner G 72, 137

H

Hager A 62, 137
Hak R 90, 137
Haldrup A 83, 108, 137
Hanba YT 98, 137
Hankamer B 13, 14, 15, 16, 29, 35, 36, 137, 138
Harbinson J 72, 79, 95
Harrer R 19, 138
Hartel H 57, 61, 138
Haussuhl K 37, 38, 138
Havaux M 3, 11, 26, 52, 56, 69, 70, 138
Havir EA 52, 56, 74, 138
Havir EA 62, 138
He WZ 17, 19, 20, 138
Heber U 71, 79, 80, 82, 139
Henmi T 18, 39, 139
Hieber AD 112, 139
Hihara Y 26, 103, 139
Hihara Y 45, 46, 139
Hihara Y 94, 96, 104
Hillier W 2, 139
Horton P 2, 3, 5, 6, 8, 47, 49, 51, 52, 53, 57, 58, 59, 65, 69, 94, 101, 114, 140
Hirata M 84, 139
Hirose T 42, 139
Hirschberg J 12, 139
Houben E 43, 140
Howitt CA 80, 140
Huber CG 7, 140
Huner NPA 3, 103, 104, 105, 140
Hurry V 26, 55, 69, 140

I

Ihalainen JA 108, 140
Ishikawa Y 39, 140

Ivanov B 75, 141

J

Jackowski G 109, 141
Jacob B 71, 141
Jahns P 34, 59, 61, 66, 141
Jansson S 2, 5, 6, 9, 10, 109, 141
Jarvis PG 104, 141
Jegerdschold C 30, 33, 142
Jensen PK 20, 108, 142
Jeong WJ 113, 142
Jin E 71, 142
Joet T 80, 142
Johnson GN 53, 80, 82, 101, 102, 103, 121
Jordon P 14, 19, 21, 22, 25
Jurik 98, 142

K

Kagawa T 48, 142
Kamite L 62
Kamiya N 13, 18, 142
Kaneko T 115, 118, 142
Kao W 48, 85, 86, 89, 92, 142
Karpinski S 3, 47, 48, 71, 75, 79, 105, 143
Kasahara M 49, 143
Kato MC 26, 33, 34, 119
Ke B 1, 29, 30, 143
Keegstra K 40, 42, 143
Kettunen R 33, 143
Khorobrykh SA 75, 144
Kim JH 41, 144
Kim JH 95, 103
Kimura M 115, 144
Kitao M 27, 47, 144
Kloppstech K 52
Knoetzel J 2, 7, 10, 144
Kogata N 42, 43
Kohorn B 2, 12, 42, 83, 114
Kok B 26, 144
Koller D 48, 84, 85, 87, 92, 120, 144
Kozaki A 80
Krause GH 27, 33, 34, 49, 53, 59, 69, 80, 81, 89, 144
Krauss N 14, 145
Krieger A 56, 145
Krol M 10, 145

Kuhlbrandt W 6, 8, 10, 13, 14, 15, 145
Kulheim C 74, 145
Kurar TA 98, 144

L

Lagoutte B 24, 145
Lancaster CRD 21, 145
Laroche J 94
Lawson T 52, 144
Lazar D 52, 146
Lee HY 34, 146
Leong SP 96, 97, 146
Levy I 94, 146
Li XP 10, 53, 54, 60, 69, 120, 146
Lichtenthaler HK 70, 90, 146
Lindahl M 36, 37, 38, 146
Lloyd JC 106
Logan A 73, 77, 146, 147
Lokstein 0, 66, 147
Long SP 1, 26, 47, 48, 51, 84, 86, 147

M

Maenpaa P 51, 147
Malkin R 17, 19, 20
Malliga P 117
Mant A 42
Marin E 70, 147
Marshall HL 35, 147
Masuda T 103, 147
Maxwell K 102
Maxwell K 53, 147
Mayfield SP 41
Medrano H 119
Mehler AH 77, 147
Meir P 104, 147
Melis A 2, 11, 12, 28, 32, 33, 51, 52, 71, 95, 103, 148
Meyer WS 72, 84, 148
Michelet B 93, 148
Miyake E 81, 119, 148
Mo F 14, 148
Morais F 43, 148
Morishige DT 2, 6, 10, 148
Muller P 47, 51, 52, 149
Mullineaux P 3, 47, 71, 79, 81, 149
Mulo P 30, 149
Munekage Y 53, 149
Murakami A 104, 149
Murchie EH 28, 94, 149
Murgia I 72, 149

N

Nanba O 17, 29, 149
Naver H 116, 149
Neubauer C 62, 149
Nield J 12, 15, 150
Nilson R 41, 43, 150
Nishikami H 72, 150
Nixon PJ 81, 150
Niyogi KK 2, 12, 47, 51, 56, 61, 69, 70, 77, 81, 84, 114, 115, 119, 150
Noctor G 58, 71, 72, 75, 77, 150

O

Obokata J 116, 151
Ogren E 28, 151
Oguchi R 84, 98, 121, 151
Ohad I 26, 33, 151
Ohnishi N 41, 151
Okamura M 119
Oosterhuis DM 87, 151
Oquist G 34, 151
Ort DR 3, 4, 47, 48, 49, 51, 53, 78, 79, 119, 151
Ortega JM 39, 151
Osmond CB 27, 50, 79, 81, 94, 95, 151
Oswald O 114, 152
Ott T 79, 152
Owens TG 2, 4, 27, 53, 54, 67, 152

P

Palatnik J 79, 152
Palmer JM 71, 152
Park YII 35. 48, 51, 152
Paulsen H 2, 4, 5, 6, 8, 152
Peltier G 116, 152
Pesaresi P 106, 115, 117, 152
Peterson RB 52, 56, 74, 153
Pfannschmidt T 12, 116, 153
Pfundel EE 68, 153
Phillip Y 66, 153

Pichersky E 2, 6, 10, 153
Pineau B 6, 153
Pogson BJ 66, 113, 153
Polle A 75, 80, 153
Powles SB 26, 154
Powles SB 85, 154
Prasad JSR 87, 88, 154
Prasil O 26, 30, 154
Psylinakis E 4, 17, 29, 154

Q

Quigg A 103, 154

R

Raines CA 106, 154
Rajendrudu G 85, 87, 88, 154
Rappaport F 12, 14, 17
Reddy VC 97, 103, 154
Rees D 57, 154
Rhee KH 13, 17, 29, 154
Richter M 57, 154
Rissler HM 113, 154
Ritter I 85, 92
Robinson C 42, 155
Rochaix JR 115, 155
Rockholm DC 62, 64, 155
Rogner M 18, 155
Rorat T 60, 155
Rose DA 1
Rosenquist E 94, 155
Rossel JB 72, 155
Ruban AV 11, 55, 58, 60, 66, 67, 68, 155, 156
Ruffle SV 109, 156
Russel WA 34, 156
Rutherford AW 32, 156
Ruuska SA 80, 156

S

Sacksteder CA 82, 156
Sailaja MV 48, 85, 87, 89, 90, 94, 96, 97, 99, 100, 101, 120, 121, 157
Sakamoto K 93, 157
Salter 102
Sandford AP 104
Santabarbara S 32, 157
Sapozhnikov DI 62, 157
Satoh K 17, 29
Sazanov LA 81, 158
Scheller MV 19, 22, 23, 24, 25, 158
Schindler C 70, 158
Schroda M 40
Sehnell DJ 42, 158
Sehreiber V 52, 158
Sehubert M 116
Settles AM 43, 158
Seuferheld MJ 54, 120
Sharma J 32, 158
Shen JR 13, 18
Sherameti I 114, 158
Sherry RA 85, 158
Shi LX 106, 107, 158
Shi LX 40, 158
Shigeoka S 73, 159
Shikanai T 52, 60, 80, 159
Simpson DV 2, 7, 10, 159
Sinclair J 35, 51, 159
Smirnoff N 72, 75, 159
Smith H 85, 159
Smith TA 42, 159
Snel JFH 52
Snyders S 2, 83, 114, 159
Somerville C 115, 158
Sonoike K 26, 45, 46, 159
Spetia C 36, 37, 38, 160
Spilotro P 58
Srivastava A 70, 160
Streb P 79, 81, 160
Styring S 30, 160
Sugiura M 42, 160
Sun WH 112, 160
Svab Z 117, 160
Swiatek M 106, 107, 108, 160

T

Tabata 115, 118
Takahashi Y 41
Takeba G 80
Tambussi EA 47, 160
Tanaka R 112, 113, 160
Telfer A 30, 32, 160, 161
Teramoto H 9, 51, 121, 161

Terashima I 26, 45, 161
Thayer 96
Thiele A 47, 69, 161
Thornber JP 8, 161
Tjus SE 26, 32, 46, 161
Tracewell CA 18, 29, 161
Trebst A 32, 162
Tremoliers A 8, 162
Tsai TT 85, 86, 92
Tsiotis G 17, 162
Tyystjarvi E 28, 32, 51, 111, 162

V

Van Kooten O 52, 162
Van Wijk KJ 40, 118, 162
Varatto C 20, 108, 162
Vasiliev S 12, 162
Vass I 30, 33, 162
Veeranjaneyulu K 12, 163
Verhoeven AS 66, 111, 112, 163
Von Caemmerer S 98, 163
Vranova E 72, 73, 163

W

Wada 49
Walker D 80, 85
Walters RG 96, 97, 101, 163
Weis E 49, 163
Weise C 49, 80, 164
Wentworth M 55, 68, 163
Werner C 27, 163
Weston E 101, 163
Wheeler GL 75, 164

Wild A 97, 164
Wilson KE 103, 164
Wingler A 81, 164

X

Xiang C 72, 164
Xu CC 73, 164

Y

Yagi K 72
Yamamato Y 39, 164, 165
Yamamoto HY 8, 33, 39, 62, 63, 64, 65, 111, 164
Yamashita N 98, 165
Yamasoto A 14, 165
Yang C 6, 103, 165
Yerkes CT 52, 59
Yin ZH 92, 93, 101, 102, 103, 121, 165
Yocum CF 17, 29
Yoder LM 17, 165
Yokota 78
Yokthongwattana K 35, 39, 165
Young AJ 27, 54, 166
Young AJ 54, 55, 65, 66, 166
Yu F 84, 86, 166

Z

Zeiger E 70
Zhang L 40, 41, 42, 43, 44, 109, 110, 166
Zheng B 37, 166
Zolla L 116, 166
Zouni A 5, 13, 16, 18, 21, 29, 106, 166

Subject Index

Acclimation response
    Changing light intensity 100-104
    Cold temperature 104
    Irradiance levels 95, 96, 103
    Limiting growth light 97
    Mature leaves 97-101
Amaranthus 98
Antenna pigments
    Core antenna 4, 5
    Organization 5, 7, 24
    Peripheral antenna 4, 5
    Size of antenna 11
Antioxidants
    Enzymatic 72
    Hydrophilic 73
    Lipophilic 73
    Non enzymatic 73
Antisense technique 113
Arabidopsis 11, 42, 57, 59, 60, 69, 72, 80, 105, 106-110, 112, 113, 115, 116, 117
Ascorbate-glutathione cycle 75
Ascorbate peroxidase 75

Barley 9
Brassica 9

CAB proteins (also see antenna) 5, 9
Capsicum annum 34
Carotenoids 8, 58
Chenopodium album 98
Chlamydomonas 9, 11, 15, 24, 35, 59, 69, 106, 115, 116, 109, 117
Chlorophyll a/b ratio 95, 96, 97
Chlorophyll protein complexes
    CP 43 14, 16, 29
    CP 47 14, 16, 29
    CP 26 59, 66
    CP 24 59, 66
    CP 29 59, 66

Chloroplast
    Genome 42
    Movements 48, 49
    Protease 37
    Proteome 116
    Serine type 38
    Stroma 28

Chloroplast recognition particle 42, 43
Chlororespiration 81
Conjugated double bonds(CDB) 60
Cosi(Cosine of incidence) 91
Cucumis 26
Cytochrome b6f complex 82
Chlrophylls – Triplet - 34
Cytochrome b 559 17, 31, 107

D1 Protein
    C-terminal fragment 35
    Deg P_2 36
    Degradation 35
    Ftsh 36
    Insertion 42-45
    N-terminal fragment 35
    Proteolysis 35-39
    Proteolytic enzymes 35-39
    Resynthesis 39-42

D2 protein 29
D-E loop 30
Deepoxidation 58, 64
Diaheliotropism 87
Digitalis purpurea 102
Dunaliella 39

Electron transport
    Cyclic around PS I 79, 80
    Cyclic around PS II 79, 80

Linear Transport
Eleusine coracana 100
Excess excitation energy 2, 3

Flourescence
    Fv/Fm ratio 89, 91, 111
    F690/F735 ratio (Laser induced) 90

Genomics
    Functional 115
    Structural 115
Glutathione 72
Glutathione reductase 73
Glycine max 86
Gomphrena globosa 100

High light stress 113

Iron-sulphur clusters 22, 23

K+ channels 93

Leaf
    Diaheliotropism 87
    Movement 85
    Paraheliotropism 86
    Solar tracking 90, 92
Leaf and chloroplast level
    adjustments for light interception 47, 48, 84
Light
    Avoidance 48, 50
    Harvesting 2-3
    LHC I 5, 25
    LHC II – Atomic model 13
    LHC II 5, 25
    LHC II major polypeptides 8
    LHC II minor polypeptides 8
    LHC, functions 11
    Maximization 84
    Quality 1, 2
    Saturation 3
    Stress 113
Lipid oxidation stress 70

Manganese cluster 16, 19
Mehler peroxidase cycle 76
MGDG 64
Microarray analysis 114

Nicotiana plumbaginifolia 67

Oxygen
    Singlet 70
Oxygen evolving complex 18, 29
ORF 62, 106

PAR 1
Paraheliotropism 86
Pea 96
Petunia 7, 10
Phaseolus vulgaris 86
Photochemical efficiency, Fv/Fm 89, 91, 111
Photochemical quenching 52
Photoinhibition
    Acceptor side 30-33
    Chronic 27
    Donor side 32-33
    Dynamic 27
Photoprotection 47
    Ascorbate – glutathione cycle 76
    Chlororespiration 81
    Photorespiration 81
    Water-water cycle 77
    Xanthophyll cycle 39, 60-64
Photorespiration 80
Photosynthetic rate 3, 97
Photosystem – I
    Reaction centre 20
    Structural organization 19-25
Photosystem – II
    Damage and repair cycle 28, 44, 45
    Low molecular weight components 16
    Photoinactivation 33, 34
    Reaction centre 17
    Structural attributes 14-19
Phototropins 93
Phylloquinone 21
Plastoquinone 114
Potato 26
PPFD 1
Proteomics 115
PsbS protein 10, 115
PsbW 15, 107
Pulvinar phototropism 92

Q cycle 31
Quenching
    Nonphotochemical (NPQ) 48, 50-54
    Photochemical (qP) 51
    qE 56, 58
    qN 57

Reactive oxygen species (ROS) 70, 72
Reverse genetic 115
Rice 7

Scavenging 71
Scotspine 10
Signal recognition particle pathways 42, 43
Singlet oxygen 34
Solanum 60, 98
Soybean 86
Spinach 18
State transition 12
Stroma 28
Superoxide dixmutase 76
Synechococcus 13, 21
Synechocystis 115, 118

Thermal dissipation of energy 49

Thermosynechococcus 13
Thylakoid associated kinases (TAKs) 12
Thylakoid membrane
    Acidifiation 61
    Lumen proteins 115
Tobacco 7
α – tocopherol 73
Tomato 7
Transcryptomics 116
Transgenics 111
Transmembrane helices 10, 16
Trans-thylakoid pH 80
Trans-thylakoid proton gradient 46

Vinca major 66
Violoxanthin deepoxidase 62-65

Water – water cycle 77

X-ray crystallography 16
Xanthophylls
    Antheraxanthin 5, 63
    Lutein 5, 13
    Violaxanthin 5, 63
    Zeaxanthin 5, 63

Ycf 9 gene 109